HOW DIVERSITY IN NATURE

In Light
of Our
Differences

AND CULTURE MAKES US HUMAN

DAVID HARMON

Smithsonian Institution Press
Washington and London

© 2002 by the Smithsonian Institution
All rights reserved

Copy editor: Jennie Reinhardt
Production editor: E. Anne Bolen
Designer: Brian Barth

Library of Congress Cataloging-in-Publication Data
Harmon, David, 1958–
    In light of our differences : how diversity in nature and culture makes us
human / David Harmon.
        p.   cm.
    Includes bibliographical references and index.
    ISBN 1-58834-066-X (alk. paper)
    1. Human beings. 2. Pluralism (Social sciences). 3. Biological diversity.
4. Human ecology—Philosophy. 5. Language attrition. 6. James, William.
1842–1910.   I. Title.
GN27.H345 2002
304.2—dc21   2002021718

British Library Cataloguing-in-Publication Data is available

Manufactured in the United States of America
09 08 07 06 05 04 03 02   5 4 3 2 1

⊚ The paper used in this publication meets the minimum requirements of
the American National Standard for Information Sciences—Permanence of
Paper for Printed Library Materials ANSI Z39.48-1984.

It is easy to perceive that the prodigious variety which appears both in the works of nature and in the actions of men, and which constitutes the greatest part of the beauty of the universe, is owing to the multitude of different ways in which its several parts are mixed with, or placed near, each other.—Jakob Bernoulli, Ars Conjectandi

Life flows on. Life ought to flow on.—Holmes Rolston III, Responsibilities to Future Generations

CONTENTS

PREFACE

Global trend watching is now an established part of political discussion, and as more and better data are gathered, predictions of what tomorrow's world will be like are steadily improving in sophistication. Still, as Octavio Paz once wrote, the future always says to us, "Not yet"; we will forever be denied certain knowledge of its details. Speaking broadly, though, it is safe to say that the unresolved social and environmental challenges of the century just past are going to have to be met, one way or the other, in that which is to come. These are the biggest of the "big issues," such as global climate change, population growth and overconsumption, the prospect of genetically engineered human germ lines, continuing economic disparity and social injustice—and destruction of the world's biological and cultural diversity. The very reason our planet can be said to be "alive" at all is because there exists here (and here alone, so far as we know) a profuse variety: of organisms, of divergent streams of human thought and behavior, and of geophysical features that provide a congenial setting for the workings of nature and culture. All three realms of difference have evolved so that they interact with and influence one another. Earth's interwoven variety—what we will call *biocultural diversity*—is nothing less than the preeminent fact of existence.

If we could simply expect this condition of difference to go on renewing itself indefinitely, then there wouldn't be much of anything to write a book about. But there is urgency in the air. According to most thoughtful scientists, biological diversity is in a state of crisis, with a sixth mass extinction in the offing unless radical changes are made in how modern Western society treats nature. There is a growing body of factual evidence and supporting theory pointing to the same kind of crisis in the realm of cultural diversity, especially betokened by the impending loss of languages and a decline in overall linguistic diversity. More and more applied work in biology, anthropology, linguistics, and allied fields is now undergirded by the assumption that we are approaching a threshold of irreversible loss, that events over the next few decades will decide whether we cross over into a fundamentally changed, less flourishing world. Such an assumption comes with a strong ethical corollary: Diversity is good, and we ought to do whatever we can to keep it.

I am convinced that the coming century truly will be one of irrevocable choices, and that one of those choices must be to preserve the vitality of bio-cultural diversity. Those convictions are the motive and justification for this book. Like any importuning author, my hope is that the reader will, in the end, join my side. Yet from the outset it must be made clear that I will not try to mount a conventional knock-out-the-opponent argument. The subject at hand rules it out, for the very quality of differentness contains more than a little ineradicable ambiguity. There must always be a measure of ambivalence and uncertainty in any fully honest appraisal of the value of diversity. For example, to leave unqualified the judgment stated above—that "diversity is good"— obviously is untenable, since diabolical varieties of suffering and evil are most assuredly part of the world's roster of differences. The task ahead is to make the case for diversity without flinching in the face of ambiguity or indulging in oversimplifications.

The aim of this book, then, is to explore the issues surrounding the mean-ing and moral imperatives of diversity in nature and culture. A summary of the overall plan may be helpful here.

To go forward, first we have to go back. Chapter 1 begins with a vignette from long ago, a memorable drawing-room encounter between two luminar-ies of the Enlightenment. It is a way to embark on our inquiry with a just sense of historical perspective, at a properly human scale. The vignette sets the stage for a discussion of one of the most persistent of all mistaken assumptions: that the unity of humankind depends on uniformity. This is an idea firmly in the mainstream of the Western tradition, emanating from the idealism of Plato. Indeed, it is more relevant than ever, for the forces of globalization are destroy-ing biocultural diversity by their incessant promotion of sameness. The diffi-culties in distinguishing unity from uniformity, which are a function of the ambiguity inherent in diversity, are treated in an extended analysis of Arthur O. Lovejoy's classic 1936 study *The Great Chain of Being*, which occupies the rest of the chapter. The focus is on a particular aspect of the Chain of Being: what Lovejoy called "the principle of plenitude," the idea that "the world is the bet-ter, the more things it contains." The self-contradictions within this principle helped hasten the collapse (at the dawn of Romanticism) of the Chain of Being as a viable explanation of the universe. Yet the "diversitarianism" of the Romantics did not bring us any closer to a modern understanding of the meaning and value of biological and cultural diversity. In the end, the Chain of Being was a failure—but a failure with an instructive outcome, one relevant to the situation facing diversity today.

The second chapter gives the essential background on that situation, that is, on the threats now facing biological and cultural diversity, and argues that the present extinction crises facing species and languages are converging to produce a wholesale loss of biocultural diversity. It opens with a discussion of the subtleties inherent in reckoning the impact of extinctions in nature and culture, and recounts the rise of the concept of biodiversity. This is followed by a summary review of biological extinctions in history and the problems in determining a background rate of species extinction and in projecting rates into the future. Next discussed is cultural diversity: what it is (and isn't) and how best to measure it globally. Language richness (the number of distinct languages in use) is, despite some inevitable shortcomings, the best proxy of global cultural diversity. This chapter draws on an analysis of the data in *Ethnologue*, the most complete, regularly updated catalogue of the world's languages now available, to review current trends in and prospects for language extinctions— and sound a cautionary note about applying the extinction concept to languages and language use. The conclusion is that current global trends threaten the "biocultural presence," a term I use to denote the entire complement of biological and cultural diversity now existing, the product of millions of years of reciprocity between organic and cultural evolution.

Chapter 3 opens out again into analytical territory through an in-depth comparison of species and languages, the keystone measurements of biological and cultural diversity, respectively. The comparison is pursued both conceptually and "on the ground." Conceptually, both species and languages prove to be very hard to define, and as a consequence the mechanisms of speciation and language genesis are the subject of continuing debate. Nevertheless, when actual global distributions of species and languages are compared, a remarkable concurrence emerges: most of the world's countries that rank highest in the number of endemic species also rank highest in endemic languages. The probable reasons for the concurrence are discussed and illustrated in various ways. The fact that species and languages are allied geographically increases the risks both face from homogenizing globalization. Still, a philosophical question remains: If species and languages cannot be defined precisely, why should we be concerned about their fate? The answer is that, despite all imprecisions, they are not simply categories in the mind; rather, they seem to correspond with real entities that function in the world. If that is so, then it implies a certain amount of imprecision being inherent in reality. Yet species and languages cohere: They are not arbitrary or chance classifications. Reconciling coherence with imprecision is the challenge. The final part of the chapter addresses this through a discussion of two opposing approaches to classification: the monothetic, or essentialist, approach, in which it is supposed that all categories can be defined by the presence or absence of definite diagnostic characters; and the polythetic approach, which says that while such diagnostic characters may not exist, coherent groups still can be formed from individuals that share a large portion of characters, even while no single character is shared by every individual. This is a critical distinction in understanding the meaning of diversity, and the discussion here calls upon philosophers (Aristotle, Whewell, Wittgenstein) and biologists (Simpson, Beckner, and Dobzhansky) to elucidate the differences. Dobzhansky's "modal points" conception of species is shown to apply as well to such components of culture as religions, languages, and ethnicity.

Chapter 3 is the theoretical heart of the book, and points us in the direction of polythetic classification as a key to understanding the meaning of diversity. Chapter 4 takes that insight and internalizes it, so to speak, by linking it, through a detailed analysis, with the psychological and philosophical thought of William James. James was one of the most important thinkers to train his sights on questions of difference. Drawing on such key works as *The Principles of Psychology, The Varieties of Religious Experience, Pragmatism*, and *A Pluralistic Universe*, a simple model of how humans distill sameness from phenomenal diversity is presented. The model validates all of the most important tenets of James's philosophy, leading to the conclusion that the nature of the universe is not that of a single monistic or absolutist entity, but instead a pluralistic (and largely polythetic) concatenation. Neither is it a deterministic universe—volition and human freedom really exist. How is this relevant today? As the field of possible experience becomes more constricted when elemental differences are erased (as through species extinctions and language loss), the human capacity to distill sameness from diversity begins to waste away, leaving us more vulnerable than ever to charismatic demagoguery, totalitarian propaganda, and disinformation, and less able to distinguish genuine diversity from sham differences foisted upon us by corporate marketing. Over the long term, this constitutes a grave threat to our essential humanity.

The final chapter considers the value of biocultural diversity. It begins with a skeptical question: Why should anyone care whether biological and cultural diversity are related or not? The reply is that seeking a holistic understanding of diversity gives us a more accurate picture of how each of us, as individuals, shares in the collective life of humankind. But another, even more serious, charge presents itself: Doesn't advocating the preservation of cultural diversity commit one to cultural relativism? We will find that the answer is "no": It is possible to support cultural diversity in general while at the same time making value judgments about cultural practices in particular. The only way to do this coherently, however, is by abandoning that search for an absolutist ethics that has so dominated Western moral philosophy. The alternative is "polythetic morality." Polythetic morality begins by recognizing the fact of moral pluralism in the world and affirming it as valuable. It emphasizes the importance of the *process of striving toward unity* rather than the actual attainment of the uniform conditions that purportedly constitute unity. This in turns leads us to a redefinition of the principle of plenitude that was discussed in Chapter 1. The new principle of plenitude combines polythetic morality with a biocultural approach to diversity preservation to state that "the world is the better for the diversity it contains." Finally, we consider whether there is an ultimate moral imperative to preserve diversity. The first key is to recognize, with James, that if diversity is the means through which our consciousness functions, and if our consciousness is what makes us human, then diversity makes us human. But we can go beyond this anthropocentric principle. In the end, the reason to preserve biocultural diversity is because to do otherwise would be to staunch the historical flow of being itself, the evolutionary processes through which the vitality of *all* life has come down to us through the ages. Once we realize this, then—and only then—can we start making genuine advances toward achieving unity among humankind and ensuring the continuation of the biocultural presence on Earth.

It is not a straight-line argument. I am going to be so bold as to claim this as an advantage, since it is so "in character" with the subject matter. I am aware that some readers would find a more synchronous and didactic presentation of the material more satisfying. Still, I feel the benefits of my approach outweigh the drawbacks. At the very least I hope those who disagree will still come away from a reading of this book with a renewed sense of the mysterious and wonderful quality of the differences in nature and culture—differences that make us human.

A NOTE ON NOTES

In the text, quotations and citations of specific ideas are attributed through author-date references. In addition, there are copious endnotes at the end of this book (some of which themselves contain quotations) that expand upon points in the text and give cross-references to other relevant sources. They are intended for readers who relish the amplifications (and yes, digressions) of notes. I myself am that kind of reader, and so find them indispensable in such a work as this. You may not. If so, that's all right. All the endnotes can be skipped without any loss of understanding.

ACKNOWLEDGMENTS

Parts of this book derive from previous publications, and I am glad to acknowledge and thank the copyright holders for allowing me to use the material here: the Linguistic Association of the Southwest (for material from articles published in *Southwest Journal of Linguistics*), the Canadian Museum of Nature (for material from an article in *Global Biodiversity*), and the Smithsonian Institution Press (for material, here radically rearranged, from a chapter in *On Biocultural Diversity: Linking Language, Knowledge, and the Environment*, edited by Luisa Maffi). Other parts are based on presentations I made at the Fourth World Congress on Parks and Protected Areas, Caracas, Venezuela, 1992, and at the following symposia: "Shift Happens: Language Loss and Public Policy," University of New Mexico, Albuquerque, 1995; "Losing Species, Languages, and Stories: Linking Cultural and Environmental Change in the Binational Southwest," Arizona–Sonora Desert Museum, Tucson, Arizona, 1996; "Endangered Languages, Endangered Knowledge, Endangered Environments," University of California, Berkeley, 1996; and "Language, Culture, and Understandings of the Environment: Lessons for Environmental Policy and Education," Northwestern University, Evanston, Illinois, and the Field Museum of Natural History, Chicago, 1999. I thank the organizers of these meetings for giving me the chance to develop some of the ideas in this book.

The best writer I know, Tom Rayfiel, poured ice water on some overheated prose in an early draft. Along the way he also challenged me on a number of debatable points; wherever I took him up, the result has been improvement. Scott Mahler provided valuable comments on the organization of the book, and was both cordial and helpful throughout the entire production process. E. Anne Bolen and Jennie Reinhardt did an excellent job of editing the manuscript. I've also been fortunate to have had Luisa Maffi, my colleague at Terralingua, to bounce ideas off of. She has been a perceptive and discerning critic, and I am grateful to her for her extremely thorough reading of the manuscript, as I am to Gary Nabhan, of the Center for Sustainable Environments at Northern Arizona University, for his. I also thank my other friends at Terralingua for all their insights as we have worked together to develop the theory and practice of protecting biocultural diversity.

To go against the mainstream, to pursue ideals that expose you to criticism rather than drift with the current, is a fundamental choice in life, and I for one could never have made it without the support of my family. My parents, James and Margaret Harmon, laid the foundation by teaching my sisters and me to always try to understand and respect the differences we find in others. Now my wife, Susan Dlutkowski, and I are trying to do the same for our two daughters. Together, we have learned a lot. I am grateful for that, and I hope much more discovery awaits us all.

VOLTAIRE'S SOLUTION

In the spring of 1765, at the height of the Enlightenment, James Boswell got his picture painted. The portrait is, more than anything, a depiction of youth. We see at half-length a man who has his life before him. From appearances he is indeed not much beyond a boy. He wears a sea-green greatcoat trimmed in the lustrous fur of a fox. Underneath, gold embroidery sets off his scarlet waistcoat. He has breeches to match. He sits, legs crossed, his right hand resting in his lap. His posture is easy and confident. His face is serious, his full, red lips pursed. He isn't handsome, but his look is striking just the same. He appears, as one scholar has written, "odd, eager, egotistical, boyish, sensual—and attractive" (Pottle 1966, 222). In this pose he manages to project precisely those qualities, faults included, by which he wishes the world to regard him. Yet the painting is also meant to be a portrait of reason in the Age of Reason: Boswell had the artist place an owl, the emblem of solemnity and wisdom, looking down on him from a perch above his right shoulder.

Sensuality and eagerness, wisdom and solemnity—these warring characteristics, captured in an arresting image from long ago, symbolize the enduring ambiguity surrounding the central question we will pursue in this book: What is the meaning and value of diversity? It is fitting to begin our search for answers on a personal level, with an impression of a specific human being in mind, because all of us have to find our own path through a world filled with difference. It is also fitting that we go back to the Enlightenment as a starting point, for perhaps no period in the history of Western thought has so thoroughly promoted the mistaken idea that the unity of humankind requires uniformity. Boswell—at that time a nonentity, but who would become the most celebrated biographer to write in English, perhaps in any language, through his unique association with the great Samuel Johnson—is the perfect embodiment of the struggle to come to terms with the implications of that idea. He spent his entire adult life in self-admonition, vainly trying to reconcile an essentially mercurial personality with the expectations of *gravitas* placed on him by the social and intellectual conventions of his time. As we begin, let us briefly follow him through a couple of remarkable encounters that will set the stage for our inquiry.

Boswell gave many precious hours during May and June of 1765 to sit for his likeness; the job admirably done, as it happened, by a fellow Scots student named George Willison who was working in Rome when Boswell stopped over as part of his grand tour of the Continent. Like most well-heeled young men of the day, he felt that such a tour was the indispensable culmination of an education. Depending on the young person's predilections, these journeys could tend toward high-mindedness or dissipation, or, as in Boswell's case, oscillate between the two. Few grand tours, though, can have equaled his for audacity. For he had come directly to Rome from France after having successfully "laid siege"—his phrase—to the two most famous authors then living: Rousseau and Voltaire.

Now, it is not unusual for a raw, unknown writer to wish to meet famous practitioners of his craft. Nor is it so unusual for such meetings to occur and even to produce a continuing relationship, should the Great Talent desire an acolyte and should the Aspirant be sufficiently interesting and complaisant. Boswell's acquaintance with Johnson had in fact gotten off much like this. Their first meeting in a London bookshop in May 1763 is one of the great literary anecdotes: Boswell making a fumbling overture, Johnson leveling the hapless Scotsman with one of his trademark ripostes (Pottle 1950, 260). Yet, characteristically, Boswell persisted and got Johnson to meet him again. Just as characteristically, he soon got Johnson to like him, and their relationship quickly matured into something far more equitable before Boswell left London behind for his two-and-a-half-year sojourn on the Continent. One reason the two got along so well was that they shared a trait of supreme importance: despite a life-long tendency to doubt, they willed themselves to an abiding respect for and faith in Christian orthodoxy.

Boswell's pursuit of Rousseau and Voltaire, which occupied the whole of December 1764, was something altogether different. His aim in meeting—in impressing—these two men was entirely self-directed. At this point in life he was acutely insecure. This was a lifelong condition with him, but at this stage of the game he had written nothing of note and thus was overwhelmed by the need to measure himself against those who most certainly had. Yes, he had quickly won Johnson's heart with his amiability and attentiveness, his uncanny ability to "tune himself" (as he said) to another's character without coming across like a bootlicker. No easy thing, given Johnson's formidable demeanor, but could Boswell do the same with authors even more renowned, more inured to flattery, who had ample skill in fending off unsolicited overtures, and who were, moreover, religious innovators of the kind he and Johnson both despised?

Of course he could. First came Rousseau. "I determined to put my real merit to the severest test," he crowed afterward to a friend, "by presenting myself without any recommendation before the wild illustrious philosopher" (Pottle 1953, 291). On December 3, 1764, Boswell lured his quarry into range with a note of introduction that concluded: "Open your door, then, Sir, to a man who dares to tell you that he deserves to enter it" (Pottle 1953, 220). Rousseau must have been intrigued by such impudence. Though in constant discomfort from a urinary tract ailment, he allowed Boswell five days at his house in Môtiers. Before it was over, they spoke seriously about literature, religion, sexuality—about every subject of importance to Boswell. They parted with affection.

From there it was on to Voltaire at Ferney. Once more, defying the odds, Boswell used his native charm to get an interview. Again he goaded the master into a freewheeling debate on matters of substance. Finally, Boswell took up the defense of orthodoxy against the razor iconoclasm of Voltaire:

> At last we came upon religion. Then he did rage. The company went to supper. Monsieur de Voltaire and I remained in the drawing-room with a great Bible before us; and if ever two mortal men disputed with vehemence, we did. Yes, upon that occasion he was one individual and I another. For a certain portion of time there was a fair opposition between Voltaire and Boswell. (Pottle 1953, 293)

"The daring bursts of his ridicule confounded my understanding," Boswell later admitted, but as the theological argument wore on, it was his host who gave ground, literally: Voltaire finally went into a mock faint, collapsing theatrically onto a chair in an attempt to shake his dogged visitor. It didn't work. Boswell placidly awaited signs of recovery, upon which he resumed the discussion in a lower key as if nothing untoward had happened. Voltaire, whose reputation as a brilliant, cutting, and often ruthless disputant was legendary (was matched, in fact, only by Johnson's ability to "toss and gore" those on the wrong side of his opinions), came, at this point, as close to being undone as he ever would. By no stretch had Boswell outshone Voltaire, but he had gone toe-to-toe with the glittering freethinker. Voltaire could not help liking him.

On December 29, 1764, Voltaire granted Boswell one last interview—for our purposes, the climactic scene. They spoke warmly and at length on weighty topics, but this time without histrionics. At last, as Boswell was getting ready to take his final leave, he confessed that, when he first came to see Voltaire a few days earlier, he was expecting to find a very great, but morally a very bad, man. Now sincerity compelled him to admit that he had been wrong: Voltaire's was a nobly good spirit, far from wicked, even if his religious principles were unsound. And yet . . . what about all those troubling, doubting articles in his newly published *Dictionnaire Philosophique*, the one on Soul, for instance? The author was diffident:

> *VOLTAIRE:* But before we say that this soul will exist, let us know what it is. I know not the cause. I cannot judge. I cannot be a juryman. . . . We are ignorant beings. We are the puppets of Providence. I am a poor Punch.
>
> *BOSWELL:* Would you have no public worship?
>
> *VOLTAIRE:* Yes, with all my heart. Let us meet four times a year in a grand temple with music, and thank God for all his gifts. There is one sun. There is one God. Let us have one religion. Then all mankind will be brethren. (Pottle 1953, 304)

With that they parted. The whole episode seems almost preposterous. It was as if a sophomore had shown up at the office door of a Nobel laureate, charmed his way in, proceeded to start (and win) an argument, and then walked out—with no hard feelings. But it happened. Boswell had a way of making such things happen. He left behind an indelible impression as he set out over the Mount Cenis Pass for Italy and, beyond, the warm, indulgent days in Rome that saw the production of his memorable portrait.

DIVERSITY, GLOBALIZATION, AND THE ROOTS OF ENDANGERMENT

I have introduced this moment from the arc of James Boswell's extraordinary life because it shows two striking personalities grappling with the meaning of diversity. It is an important issue, because diversity—the range of differences found in nature, in culture, or at the boundary where the two meet—is imperiled today as never before. As it never *could have been* before, for today's particular confluence of technological advancement and post-Cold War political rupture has set up an unprecedented situation, which, for better or for worse, has come to be known under the rather loose term *globalization*. As we shall soon see, the roots of globalization reach deep into the nourishing soil of Western thought. But the socioeconomic forces driving current trends have only just begun to move at a speed that outstrips the ability of traditional social responses to cope, and are only now attaining a truly global reach.

I should make it clear at the outset that this book is not meant to be another critique of globalization; there are already plenty of those. The main question, again, is the significance and worth of diversity. Globalization imparts an insistent urgency to the question, but seeking an answer would be important even if we were not faced with the prospect of massive extinctions in species and languages, the two main indicators of biological and cultural diversity, respectively. The brute fact of diversity in nature and culture—that things *are* different, indeed luxuriously different—is a fundamental characteristic of life that has gone largely unexamined, or at least has been probed only peripherally, outside of a small circle of professional philosophers. And even less well-understood are the connections between biological and cultural diversity. Too often they are considered to be on entirely separate public-policy tracks—a lamentable state of affairs, since this false division precludes an integrated response to preserving the vitality of the commingled evolutionary processes that produced the diversity in the first place. More importantly, it prevents us from seeing how diversity is essential to our humanity.

Now, returning to our introductory scene at Ferney, if we are to make sense of Boswell and Voltaire's exchange within the context of the issues just outlined, we must first of all understand that they were not simply debating religious orthodoxy. As Voltaire's final remark shows, religion is to be regarded as a solution to a worldly problem: that of the elusive unity of humankind. The cause of the problem, we can infer, is the disconcerting gulf between the variety that characterizes existence and the ideal of unity, that vision of a far-off horizon toward which all right-thinking people are supposed to turn their eyes.

That the problem and its cause are not explicitly stated is telling in itself: Voltaire could count on Boswell to agree with his premises. This is not surprising. The rhetorical chain "one sun, one God, one religion, one people" is forged with a logic that has proven irresistible throughout the history of Western thought. It is a logic that says that, somehow, there must be an eternal world of Oneness where God and harmony reign; that the visible world of everyday life with all its messy and incorrigible differences is either a pale copy of this real world or else an outright illusion; that earthly diversity therefore must be subordinated to the ideal of heavenly harmony, for, to the extent that it cannot be, it becomes something positively evil because it distracts and separates us from the Oneness, the Absolute, which is the highest good; and that, therefore, a primary objective of human affairs should be to overcome differences by subjugating or, if necessary, eliminating them.[1]

At least part of this reasoning can be traced to the idealism of Plato, whose influence helps explain why it endures. And it thrives today, a perennial reaction against the seeming impossibility of ever grasping this unruly world of ours by the shoulders, giving it a good shake, staring it down, *knowing* it all at once. For even today in the vaunted global village there are still faraway places, unimaginably separate; places dimly populated by strange plants and animals and peoples, none of which look like Home. How does it all make sense? Or does it make sense at all? The psychologist and philosopher William James— a touchstone later in this book—felt the importance of this all-too-human bewilderment. Whatever the "real world" is, he considered, common sense tells us it must consist of "the sum total of all its beings and events" as they have unfolded throughout history and continue to unfold right now.

> But can we think of such a sum? Can we realize for an instant what a cross-section of all existence at a definite point in time would be? While I talk and the flies buzz, a sea-gull catches a fish at the mouth of the Amazon, a tree falls in the Adirondack wilderness, a man sneezes in Germany, a horse dies in Tartary, and twins are born in France. What does that mean? Does the contemporaneity of these events with each other and with a million more as disjointed as they form a rational bond between them, and unite them into anything that means for us a world? (James 1890, 2:635)[2]

Over the centuries, the dominant Western philosophies and theologies have answered these questions negatively, choosing instead to follow the chain of logic described above to its otherworldly terminus (Lovejoy 1936, 25–27). James spent his whole life fighting this kind of absolutism, though he was fully aware of its popular appeal and not only recognized but ardently defended the psychological impetus behind it: the need to seek mastery over despair and confusion.[3] What he despised was the wholesale deferral of hope and clarity to an unseen world-in-waiting.[4] In a famous passage he declared, "If this life be not a real fight, in which something is eternally gained for the universe by success, it is no better than a game of private theatricals from which one may withdraw at will. But it *feels* like a real fight" (James 1897, 61). James was a pluralist, a believer in the reality of the diversity in the here-and-now world, the world of our senses, the mundane world where all of us eat our lunch, whatever our doctrines may be. James' philosophy, taken as a whole, is a vigorous counterargument to absolutism, and we shall explore it further. For now let us simply note that his has been a decidedly minority view.

So the possibility Voltaire declaimed so long ago—that of perfecting the world by making the key elements of everyone's life experiences substantially the same—is firmly in the mainstream of Western thought. In 1764 the possibility was merely an interesting speculation; today, it is fully alive, up for grabs, and central to the planet's future. In the short time since the end of the Cold War, the homogenizing socioeconomic and environmental forces of globalization have begun to converge with astonishing power. Faced with an unprecedented cross-reinforcing process in which the planet's biological and cultural diversity is being destroyed, we have now to consider whether we might actually get the kind of world the two great writers were talking about—and soon. For the first time in history, Voltaire's solution is within our grasp.

Much of what is currently written about globalization treats it as a phenomenon *de novo*, a confluence of high tech and recent geopolitical shifts. But the exchange between Boswell and Voltaire shows us that the forces behind it are not just newly unfettered expressions of a mercenary materialism running amok on a post-Communist, uplinked, hypercorporate planet. The pace and scope of globalization are truly novel, but its lineage is venerable, far predating even the Enlightenment. None of the overt forces of globalization we have so quickly become used to—the worldwide overwash of American TV, the concentration of wealth and power in gigantic transnational businesses, the

concomitant decline in the primacy of national governments, and so forth—
could have come about were they not firmly laid on a long-standing foundation of thought and sensibility which, unwittingly or not, devalues the integrity
of Earth and earthly existence. By "integrity" I mean the entire complement
of biological and cultural variety as it has evolved over the millennia, and which
(as I will argue) has melded into an integral, indissoluble biocultural presence
whose existence has been, and remains, a crucial part of what it means to be
human. Globalization, by emphasizing homogeneity, speed, instant gratifica-
tion, packaged experiences, and all the rest, trivializes this biocultural presence
and, in so doing, threatens to extinguish it. This change in the course of the
biosphere and of civilization did not suddenly emerge along with cell phones
and the fall of the Berlin Wall. The groundwork was laid by centuries of
thought that stressed that the promise of existence in a kingdom-to-come is
more real and valuable than the diversity of life on Earth.

Thus the most important trend of our time can lay claim to a philosophical and intellectual imprimatur dating back to the beginnings of Western civilization. This fact transforms the absolutist-versus-pluralist debate from a dry
exercise in metaphysics to a compelling political issue, for if we are to understand globalization, we must understand its foundations. Obviously, the particular world view that has made globalization possible—that diversity produces discord, and uniformity is the only road to unity—is extremely powerful
and has long since suffused the public consciousness.[5] There is no mystery as
to why. To value diversity at all, one first must be able to appreciatively enter
into a very wide range of thought and feeling. Yet this is not nearly enough. To
value diversity fully, one cannot just cultivate tolerance. It takes "imaginative
insight into the points of view, the valuations, the tastes, the subjective experiences, of others; and this not only as a means to the enrichment of one's own
inner life, but also as a recognition of the objective validity of diversities of valuation" (Lovejoy 1936, 304). Anything less can all too easily lapse into a shallow
eclecticism—or, worse yet, a self-referential exoticism.[6] Meanwhile, unity is
unity is unity. As an ideal, it is easy to understand. It inspires us. It gives us
things like the finale of Beethoven's Ninth Symphony.

To return once more, then, to the question raised by Voltaire's exchange
with Boswell, I suspect that the answer most people will give is, yes, we would
be better off if everyone's life experiences were pretty much the same. In fact,
the answer appears self-evident: Harmony is better than discord, agreement

better than disagreement, peace better than strife. What sane person could argue otherwise?

Voltaire himself did, as a matter of fact. Admittedly, two of the (very few) consistencies of his long and turbulent life were the profession of deism and a belief in the existence of a natural moral law applicable to everyone; these at the very least point toward the possibility of universality in religious experience. Yet his attitude toward the concept of a supreme deity was to posit God as a sheer political expediency, a necessity for good government and civil order (Besterman 1969, 220–223). The tenor of his last meeting with Boswell suggests this, for surely his self-effacing reply to the question about the Soul was an affectation. Voltaire was no "poor Punch," and he of all people knew it. Add to this the fact that he was an implacable enemy of religious intolerance, a stance that had cost him dearly more than once over the years. Voltaire was thus better placed than most to judge whether the imposition of a single set of beliefs would, in real life, induce all men to live as brothers. His personal experience told him that, if nothing else, the champions of the dominant religion in France were willing to commit all sorts of outrages to bring heretics to heel. This man—the author of *Candide*, the most devastating satire on fatuous optimism in all of literature—knew in his bones how dangerous simplistic formulas for social harmony can be.[7]

So was Voltaire in earnest in his final words to Boswell? Or was what I have called his "solution" just a blow-off line, a way to hustle his unbidden visitor out the door? We don't know; the signs point in both directions at once. And that in itself is a crucial point to understand. Ambiguity, it has been said, is systemic in the universe (Hick 1989, 124). Nowhere is this more true than in questions surrounding the meaning and value of diversity. Those who demand cut-and-dried, either-or answers will be disappointed. This is not to say that there are no answers at all, only that the answers are nuanced and the path to them is sometimes long and arduous. No shortcuts will suffice.

THE GREAT CHAIN OF BEING AND THE PRINCIPLE OF PLENITUDE

Consider, as an extended example, the place of diversity in the history of one of the most influential ideas in all of Western civilization, the Great Chain of Being, as elucidated by Arthur O. Lovejoy in his classic 1936 study of the same name.[8]

The Chain of Being was the dominant metaphysical image employed in the West before the Darwinian revolution. It pictured a universe where each kind of being from the "lowest" organism on up to angels at the foot of the Throne of Heaven had its appointed place in the cosmic hierarchy. The degrees between each station were, to the mortal mind's eye, almost infinitesimally fine, their discernment the province of God alone. In this divine exclusivity the Chain of Being ran the risk of being unfathomable. Left unremedied, such inscrutability would have been fatal to it as a popularly accepted image. But it was saved by three additional qualities. First, it was ordered: As the metaphor of the chain denotes, it was a linear hierarchy leading up to God. And it was a chain laid straight: There were no outliers, no oddball cases out of the line of perfection. Every form of being, no matter how bizarre, was understood to fit somewhere along the line. Second, the places in the hierarchy were appointed. No kind of happenstance in the ordering was admissible. The result was the grandeur of inexorability, an awesome evidentiary statement of the majesty and mastery of the Creator. Third, each species of being in the hierarchy possessed an immutable essence. Each was what it was meant to be, just as it had always been since the moment God saw fit to bring it into existence, and it was going to be just that and nothing else until the end of time. Although outward appearances may vary from individual to individual, there was no question of whole new forms of life coming into being or departing, of speciation and extinction in an evolutionary sense.[9] Metamorphosis and transition were surficial. This made for an architecture of the universe deeply satisfying to generation after generation, and not just to the simple faithful, but to the most penetrating intellects the Western tradition could produce.

As Lovejoy explained it, the Great Chain of Being is based on three related principles, all of which had sources in Plato. The first two—continuity, the idea that each kind of being differs from the next only by the smallest of degrees, and gradation, the idea that these beings form a hierarchy graded from lowest to highest—will concern us no further. The third is what brings in consideration of diversity. Starting from a complex textual analysis of the dialogue *Timaeus* (in which Plato considers the structure of the physical world), Lovejoy argued that the Chain of Being also contains within itself a countervailing principle. This "principle of plenitude" is difficult to paraphrase because it rests on a "bold logical inversion": The true attainment of good, which is mandated by the necessary existence of a perfect Absolute (often thought of as God) at the top of the Chain of Being, can't happen unless the universe is like a plenum in which "the range of conceivable diversity" is "exhaustively exemplified" (Lovejoy 1936, 52). Why? Because a totally complete and self-sufficient being "whose perfection is beyond all possibility of enhancement or diminution" could not possibly be envious of things outside of itself. Moreover, if this supreme being did *not* produce these other things (though imperfect in themselves), by this omission "it would lack a positive element of perfection, would not be so complete as its very definition implies that it is." Hence the "timeless and incorporeal One" becomes "the logical ground as well as the dynamic source of the existence of a temporal and material and extremely multiple and variegated universe" (Lovejoy 1936, 49).

This is, to be sure, an abstruse beginning—the more so since, if Lovejoy's interpretation of the *Timaeus* is correct, Plato is there contradicting his famous cave analogy from the *Republic*, in which what we commonly call "the real world" is likened to mere shadows on a cave wall thrown by eternally existing (yet unseen) Forms that constitute true reality.[10] Be that as it may, Lovejoy carries forth the argument thoroughly and clearly, following all the way down to the German Romantics the idea that "the extent and abundance of the creation must be as great as the possibility of existence and commensurate with the productive capacity of a 'perfect' and inexhaustible Source, and the world is the better, the more things it contains" (Lovejoy 1936, 52). The starting point is the observation of diversity around us ("extent and abundance of creation"), whose potential breadth is asserted to be of necessity ("must be as great as the possibility of existence") *because* it issues from and reflects God; so—and here is the ethical turn—a maximally diverse world is not just good, but the best possible world.

This is giddy stuff. But immediately it presents a seemingly devastating contradiction, for if every possible niche of creation is filled, the bad as well as the good will be endlessly exemplified. Far from running away from this conclusion (which is an extreme statement of the famous "problem of evil"), the intellectual heirs of Plato actually embraced it. Plotinus, writing in the third century, was one of the first. Here is Lovejoy, paraphrasing:

> The optimistic formula itself, in which Voltaire was to find the theme of his irony in Candide, was Plotinian; and the reason which Plotinus gives for holding this to be the best possible world is that it is "full"—"the whole earth is full of a diversity of living things, mortal and immortal, and replete with them up to the very heavens." Those who suppose that the world might have been better fashioned do so because

they fail to see that the best world must contain all possible evil. . . . Conflict in general, adds Plotinus, is only a special case and a necessary implicate of diversity; "difference carried to its maximum *is* opposition" (Lovejoy 1936, 64, 65–66).[11]

How can a God-perfect world contain evil and conflict? Why *must* it? To explain, theologians (and allied philosophers who defend traditional theism) have fashioned theodicies—formal justifications of the existence of evil, which is itself explained as being a spur to greater goods that would not exist otherwise.[12] Viewed sympathetically, theodicies are delicate reconciliations, invitations to forbearance that allow faith to remain intact in the face of the cruelest turn of fortune. Viewed unsympathetically, they are oversubtle and empty, puny in proportion to the amount of evil they are supposed to offset. Viewed cynically, they are a ploy to dupe the faithful, a way of sugarcoating the unpalatable truth. Whichever view one takes, the important point here is that the palpable fact of earthly diversity had to be accounted for in the Great Chain of Being, and in so doing its proponents were forced to deal with the problem of evil. The principle of plenitude was the result, and it in turn transformed the Chain of Being from an explanatory metaphor of the universe's structure into a theodicy that explained much more.

The upshot was that diversity became associated with the logical possibility of evil, even while at the same time it had to be exalted as a product of God. And there was a corollary self-contradiction: as Lovejoy shrewdly noted, both the later Platonists and medieval clerics "were equally committed to the two contradictory theses that 'this' world is an essentially evil thing to be escaped from, and that its existence, with precisely the attributes it has, is a good so great that in the production of it the divinest of all the attributes of the deity was manifested" (Lovejoy 1936, 96).

Self-contradictory or not, theologians, philosophers, and popular writers alike found the principle of plenitude necessary precisely because it was necessary to the Great Chain of Being—and the alternative to *that* was unimaginable. So plenitude found an exponent in writers as diverse as Thomas Aquinas,[13] Gottfried Wilhelm von Leibniz (whose only book on philosophy published during his lifetime was titled *The Theodicy*), the essayist Joseph Addison, and the Anglican divine Edmund Law.[14] There were a host of others.

The Chain of Being found its widest acceptance during the Age of Enlightenment (Lovejoy 1936, 183). Continuous, complete, rigid, and unchanging, it accorded perfectly with ideals whose "ruling assumption was that Reason—

usually conceived as summed up in the knowledge of a few simple and self-evident truths—is the same in all men and equally possessed by all; that this common reason should be the guide of life; and therefore that universal and equal intelligibility, universal acceptability, and even universal familiarity, to all normal members of the human species, regardless of differences of time, place, race, and individual propensities and endowments, constitute the decisive criterion of validity or of worth in all matters of vital human concernment . . ." (Lovejoy 1936, 288–289).

> The Enlightenment was, in short, an age devoted, at least in its dominant tendency, to the simplification and standardization of thought and life—to their standardization by means of their simplification. Spinoza summed it up in a remark reported by one of his early biographers: "The purpose of Nature is to make men uniform, as children of a common mother."[15] The struggle to realize this supposed purpose of nature, the general attack on the *differentness* of men and their opinions and valuations and institutions—this, with the resistances to it and the eventual revulsion against it, was the central and dominating fact in the intellectual history of Europe from the late sixteenth to the late eighteenth century. (Lovejoy 1936, 292)

In its *dominant* tendency, yes. But there are always other tendencies: the resistances, and, eventually, revulsion. As a program, the Great Chain of Being was eminently suited to the Age of Reason. Yet the application of reason proved to be its undoing. By the early eighteenth century enough solid evidence of fossils had accumulated to demonstrate that extinction was not only conceivable, but a regular feature of history. It is an ironic fact that, before it could grasp the mechanism of life as revealed by Darwin, Western thought first had to pass through the portal of death, as it were; for only by first admitting the reality of extinction could truly scientific theories of evolution and speciation develop (Eiseley 1958, 53–54). As early as 1713 the great English botanist John Ray was at the threshold of the door, boldly declaring that "many species of Shell-Fish are lost out of the World, which Philosophers hitherto have been unwilling to admit, esteeming the Destruction of any one Species a dismembring of the Universe, and rendring it imperfect; whereas they think the Divine Providence is especially concerned to secure and preserve the Works of Creation" (Lovejoy 1936, 243).[16] Ray certainly was not out to upset all the conventional notions of Divine Providence. *On the Origin of Species* lay a century and a half down the road, so for the time being the Chain of Being's appointedness (i.e., the doctrine that each species was a special creation of God) remained effectively unchallenged. Clearly, though, the proof of extinction meant that the Chain's immutability and orderliness had taken a fatal hit.

Once these contradictions to the whole scheme had been discovered, the principle of plenitude grew to appear much less necessary. Its partisans did not give up without a struggle,[17] but by the time of Boswell's conversation with Voltaire the fight was just about over. In fact, Lovejoy credits Voltaire himself with pointing out the root defect of the principle of plenitude: he saw that "it left no room for hope, at least for the world in general or for mankind as a whole. If all partial evils are required for the universal good, and if the universe is and always has been perfectly good, we cannot expect that any of the partial evils will disappear" (Lovejoy 1936, 245). Issuing from otherworldly, absolutist assumptions, the principle of plenitude grated on that portion of human nature that craves potentiality over certainty. Though it took centuries for the dissonance to come to the surface, once it did the principle of plenitude was seen to violate its own pretensions to completeness. It is no small irony that this dissonance, which ran counter to the Enlightenment program, was borne up from the depths by the very science the Enlightenment so exalted.[18]

THE ROMANTIC TRANSFORMATION AND BEYOND

The end—the eventual revulsion—finally came in the form of Romanticism. If the Enlightenment was a movement of science and philosophy, the Romantic period was one of the arts. Lovejoy gives a catalogue of the results, including the "immense multiplication of genres and of verse-forms," the admission to aesthetic legitimacy of mixed genres, the "quest for local color," the attempt to "reconstruct in imagination the distinctive inner life of peoples remote in time or space or cultural condition," the revolt against simplicity as an aesthetic, the "distrust of universal formulas in politics," and the antipathy to standardization in any artistic endeavor (Lovejoy 1936, 293).

The Chain of Being had finally broken down "largely from its own weight," and the static principle of plenitude gave way to a dynamic world view of increasing diversity. The God whose attributes were thought to have been revealed by the Chain of Being was now seen to be "one who manifests himself through change and becoming"; the tendency of nature, to produce new kinds; the destiny of the individual, to be free to grow personally (Lovejoy 1936, 245, 296).

The "destiny of the individual"—here is the heart of Romanticism. In the universe pictured by the Chain of Being, there was little or no scope for individuality. One's person is subsumed into the mass of humankind, and humankind is subsumed into the Chain.[19] Such a suffocating determinism finally had become intolerable. In its place bloomed the unforgettable hero as envisioned by the poet Byron: a brooding, great-souled individual, struggling alone to remake the world as if through an act of will, destined to meet a tragic fate. When a personage is as intensely generative as this, the variety in nature and culture becomes an immediate, dynamic source of inspiration rather than the emblem of a distant, changeless creation that took place at the dawn of time. The Romantic temperament took the principle of plenitude, disengaged it from the ruins of Platonic metaphysics, individualized it, and finally exalted it.

Lovejoy gave the name "diversitarianism" to the transformation, and he traced its course by following the German strain of Romanticism, starting from its progenitor Friedrich von Schiller and working through Wilhelm Wackenroder, the Schlegel brothers August and Friedrich, and the theologian Friedrich Schleiermacher. He made no attempt to survey the other, very different strains that emerged in England, France, America, Russia, and elsewhere. For the purpose of our discussion it is just as well, because we need only one quote from Schleiermacher to illustrate how quickly the tide had turned. Recall that the conversation that produced what we have been calling "Voltaire's solution" took place at the end of 1764. Just thirty-five years later, Schleiermacher was declaring the opposite: "Why, in the province of morals, does this pitiable uniformity prevail, which seeks to bring the highest life within the compass of a single lifeless formula?"

> The different manifestations of religion cannot be mere subdivisions, differing only in numbers and size, and forming, when united, a uniform whole. In that case every one by a natural progress would come to be like his neighbor. . . . I therefore find the multiplicity of religions to be grounded in the nature of religion. . . .[20] You must abandon the vain and foolish wish that there should be only one religion . . . for no one will have his own true and right religion if it is the same for all (Lovejoy 1936, 308, 310–311).[21]

Within a generation, centuries of certitude evaporated. The Great Chain of Being was finished as a viable explanation of the universe, and the sovereignty of Reason was impugned. Thus cleared, the stage was open for new alternatives.

which diversity is only a device to explain away evil; to the Enlightenment, in which diversity is an impediment to harmony; to Romanticism, in which diversity is a mere apparatus for artistic egotism.

A failure, then; or perhaps, more precisely and justly, a failed experiment with an instructive outcome. In the end, this was the view to which Lovejoy inclined. "The discovery of the intrinsic worth of diversity," he declared, "was . . . with all the perils latent in it, one of the great discoveries of the human mind; and the fact that it, like so many of his other discoveries, has been turned by man to ruinous uses, is no evidence that it is in itself without value. In so far as it was historically due to the age-long influence, culminating in the eighteenth century, of the principle of plenitude, we may set it down among the most important and potentially the most benign of the manifold consequences of that influence" (Lovejoy 1936, 313). Despite all its failings, perhaps there is something to salvage from the principle of plenitude after all. If so, then it will have to be redefined to make it relevant to what we now understand about the world and what is happening to its biocultural presence. That is what I attempt in the final chapter.

But first we must look more closely at the current situation of biological and cultural diversity. In briefest terms, here are the two critical trends: (1) we have embarked upon the sixth mass extinction of species in history, the first to be caused by humans (Groombridge 1992; Wilson 1992; Heywood 1995); and (2) we are also poised on the edge of a century that will very likely see a mass extinction of languages, something that probably has never happened before in such a short time (Krauss 1992, 1995a; Harmon 1995). For a variety of reasons, the conventional approach to the two extinction crises has been to treat them separately and in isolation from each other. That is beginning to change. A new wave of interdisciplinary studies is considering how these two realms of difference are related and what common factors are at work to diminish them; call it the dawning of a biocultural approach to diversity. My central argument is that the wholesale loss of biocultural diversity, as measured through the mass extinction of species and languages, is nothing less than a threat to our common humanity. Let us turn next to laying the groundwork for that claim.

THE CONVERGING
EXTINCTION CRISES

The failure of the Great Chain of Being to lead to a modern understanding of biological and cultural diversity can be read as an indictment of one of the inadequacies of Western thought prior to the twentieth century: namely, that it consistently failed to explain the "overlappingness" of life. The whole series of demarcations that made up the Chain were far too dogmatically discrete to accurately portray the complexities of either the natural world or human civilization. If the intellectual enterprise of the last one hundred years teaches us anything, it is that the journey to a full and proper explanation of complex phenomena can begin only after one abandons a restricted point of view. Pursuing cross-fertilization among different disciplines, with their radically disparate techniques and stores of knowledge, obviously is an arduous task. The going is slow. We are only now in the earliest stages of understanding the global dimensions and ramifications of diversity in nature and culture. What has been learned so far has come largely from separate lines of investigation: biologists focusing exclusively on biological diversity, and anthropologists and linguists focusing on cultural diversity.

Yet there are many ways in which biological and cultural diversity influence and even interpenetrate one another. This is where cross-fertilization among modes of inquiry will pay dividends. In fact, it already has—most emphatically in the increasing realization, among scholars and the general public alike, that we are on the verge of not one, but two mass extinctions: of species and of languages. Authors have produced a rising stream of articles in academic journals and the popular press, and a few are even beginning to explore the possibility that the two forms of extinction are connected.

Heightened intellectual awareness of the major trends affecting the biocultural presence is a welcome development. Still, it is all too easy for the impending loss of diversity to remain a distant, abstract, even arid threat when considered as a worldwide phenomenon. For many people, emotional involvement begins by confronting the issue on a personal scale. The seeds of caring flower first in places people know by heart. Then—one hopes—the circle of concern expands.

Scientists are not exempt from love. In 1992, the biologist Edward O. Wilson published a remarkable book called *The Diversity of Life*. In it he summarizes what Western science knows about the "wondrous diversity of life that still surrounds us" and how humans have been shaped by it. One of his stories describes some rare plant species in the Andean foothills of Ecuador that were devastated when the ridge on which they grew, called Centinela, was cleared for planting cacao and other cash crops. As it happened, two botanists had begun studying the ridge some years before; otherwise the species would have lived and died unrecorded in the formal literature. Most species extinctions in the years to come will go entirely unwitnessed by academic scientists—"not open wounds for all to see and rush to stanch but unfelt internal events, leakages from vital tissue out of sight" (Wilson 1992, 243). The fight to save the California condor is very much the exception, not the rule. "Any number of rare local species," says Wilson, "are disappearing just beyond the edge of our attention. They enter oblivion like the dead of [Thomas] Gray's *Elegy* [*Elegy Written in a Country Church Yard*], leaving at most a name, a fading echo in a far corner of the world, their genius unused" (Wilson 1992, 244).

To me, to anyone who shares Wilson's feeling for nature as a source of awe, this is a poignant lament. I have never seen Centinela and in all likelihood never will, but it doesn't matter. I am a conservationist by trade—by calling, actually—and Wilson has captured the desolate feeling that (I strongly suspect) all of us in the conservation professions succumb to when we hear the latest news of yet another setback. In this impotent mood, we feel the richness of Earth slipping away despite all our efforts to the contrary. Our impotence is compounded by frustration: We cannot fathom how anyone could fail to share our concern, to be moved as we are by the beauty—the moral authority, even—of the diversity of nature.

And yet, a close examination of Wilson's quote reveals that there is no high road to such dreams. He slips easily, invitingly, into the use of the universal pronoun in the phrase "just beyond the edge of *our* attention." But exactly whose consciousness, whose attention, were these rare plant species "just beyond"? Could it be that these plants were *not* beyond the consciousness of local people—perhaps the very same farmers who cleared the ridge? Could it be that they knew the plants—maybe not as well as their grandparents had, but still knew them—yet went ahead and plowed them under? Could the destruction of these species be a perverse consequence of an earlier loss of diversity: a diversity of consciousness, of culture, that valued the whole complex of wild plants in situ more than the cash crops that replaced them? A culture that did, in fact, "use the genius" of the rare plants, perhaps as food, as sacrament, as . . . medicine?

> I turned to a young man who was carrying in his arms his two-year-old daughter suffering from diarrhea; he had already struck me by telling me he had started out at dawn from his isolated household to get to the clinic—hours of walking, and now hours of waiting, with his sick child in his arms, hours of delay in getting treatment, a delay that might well prove fatal to her. With mounting anguish I asked him whether he knew of any plants or other local remedies for diarrhea, even if he had not tried to administer them to his daughter. He searched his mind, apparently in vain, then looked to another, slightly older man nearby, and started an animated discussion in Tzeltal with him. It became clear that between the two of them they were trying to dredge up and piece together scattered fragments of latent ethnomedical knowledge—knowledge perhaps only imperfectly learned, never concretely used, and now almost forgotten. I heard them question each other: 'What's its name, the grasshopper thing?' The 'grasshopper thing': *yakan k'ulub wamal* 'grasshopper leg herb' (*Verbena litoralis*), one of the commonest diarrhea remedies in the Highlands. They could hardly remember its name, let alone master its use. (Maffi 2001a, 2–3)

In this part of the Mexican state of Chiapas, the linguistic anthropologist Luisa Maffi goes on to tell, *yakan k'ulub wamal* grows everywhere—probably it was right outside the young man's back door. Moreover, like many other traditional Mayan botanical medicines, it is no mere nostrum but often an effective

remedy. Why hadn't he used it? Perhaps, Maffi speculates, "he could not recognize it, or if he did, he clearly did not know how to use it. Or maybe he had
actually pulled it out as a weed, as my own collaborator had told me he had
unwittingly done with medicinal plants his late father, a traditional healer, once
kept in his house garden. . . ." The fate waiting for this young father and his
daughter was to try to get what care they could from the visiting trainee of the
Mexican Health Services—care which, in Maffi's experience, would likely be
not only unsympathetic and superficial, but culturally, even medically, inap-
propriate (Maffi 2001a, 3).

Stories such as these are emblems, at a human scale—within the first circle
of caring—of the ongoing worldwide loss of biological and cultural diversity.
They are stories of loss, to be sure, but they are also much more. They are
parables of existence and nonexistence. They demonstrate how reality is par-
titioned into spheres of knowledge by different groups of people.

Consider this: What if the two botanists in Peru had *not* reached Centinela
before it was cleared? The plant species lost there would have remained invisible within the sphere of knowledge called "Western science." In fact, as
far as scientists are concerned, it would have been as if the species never existed.
When undiscovered species become extinct, their existence is not merely
erased. It is negated.

But, as I have already surmised, it is probable that the Centinelan species
were not wholly undiscovered. Ethnobotanical research has shown that species
need not even be considered useful to be recognized within systems of "traditional environmental knowledge"; sometimes it is enough for a species to be
considered merely distinctive. It is therefore likely that long before the scientists' arrival, the existence of the species would have been affirmed by people
living nearby. Affirmed, yes, but later devalued, just as the whole sphere of traditional environmental knowledge about Centinela and places like it is shrink-
ing (Hunn 2001; Zent 2001). This largely unseen and unregarded "extinction of
experience" (Nabhan and St. Antoine 1993, 229–250) is both an erasure and
an outright negation of existence.

Shifting back to the scene in Chiapas, we see now how the extinction of
experience can leave a little girl's life hanging in the balance. What about her
existence? Just what is at stake here? Of course, if she died there would have
been the immediate emotional devastation visited upon her family and friends.
But, more than that, who can reckon the loss of *potentiality* brought on by a
premature death?[1] Again, this kind of loss does more than just wipe the slate

of existence clean: it ensures than nothing new will get written in its place. Something incipiently real is thus needlessly, wastefully consigned to oblivion, to nonexistence.

Such are the subtleties in reckoning the impact of extinctions in nature and culture. It is not like keeping a simple two-column ledger with debits and credits. Biologists and ecologists discuss habitat conversion and fragmentation, invasions of exotic species, global climate change, and the other factors that erode biological diversity at all its levels, from loss of genetic variation to species extinctions to the impairment and destruction of ecosystems. They recognize, of course, that everything is being driven by human social forces. Likewise, anthropologists, sociologists, linguists, and other social scientists discuss cultural disruption, language endangerment, social disaffection, globalization, and the other factors that go into the loss of cultural diversity at all its levels, from the psychology of the individual to large-scale ethnic, cultural, and linguistic groupings. They in turn recognize that the impoverishment of nature plays an important role in how cultures around the world are being transformed, and they have been quicker than biologists to note the parallels between the loss of biological and cultural diversity.[2] Even so, an integrated approach to diversity is only now beginning to take form along the frontier where the biological and social sciences meet. We see it particularly flourishing in the new hybrid fields of ethnobiology and environmental anthropology, with resonances in historical ecology, environmental ethics, ecological economics, and a host of others. It is here that systematic attempts are being made to analyze the interaction of ecological and social factors as they affect diversity. In this perspective, biological and cultural diversity are not separate: they penetrate one another.[3] In fact, perhaps the main lesson so far from these new fields is that often the two *permeate* each other.[4] Hence the increasing use of the term "biocultural diversity."

BIODIVERSITY AS SCIENTIFIC CONCEPT AND RALLYING CRY

At this point, however, let us examine the word "biodiversity," another new term that has gained even more currency. It was coined in 1985 by Arthur G. Rosen at the first planning meeting for the U.S. National Academy of Science's "National Forum on BioDiversity" (so spelled), which took place in September 1986 (Heywood 1995, n. 2; Takacs 1996, 37–39). An unusually high-profile conference, it drew not only top scientists but also reporters from leading American news outlets.[5] The papers from the meeting were published under the editorship of Wilson and became an academic bestseller (Wilson 1988a). The same period saw the rise of conservation biology, a self-proclaimed "crisis discipline" (Soulé 1985, 727) which (unabashedly) puts scientific research in the service of environmental advocacy, with a primary focus on biodiversity.[6] It quickly became one of the fastest growing academic fields.

Why this remarkable success? Why the need for a new word and a new discipline, and why now? After all, fascination with natural variety is as old as speculation itself. Every culture has its creation story, a narrative template that not only explains the group's significance (and often asserts its centrality) but also physically locates it within nature and nature's fecundity. A not-too-dissimilar analytical thread runs throughout Western scientific thought, and if we are looking for its apotheosis, all we have to do is say "Darwin." Although many scholars have pointed out that he pretty much failed to explain the origin of species (e.g., Eiseley 1958), the fact that he so titled his great work attests to the importance of the question. Explaining the palpable variety of nature is one of the perennial challenges existence presents to the human intellect.

The biodiversity movement—and it is a movement—has been successful because it concentrates the long-standing, but heretofore generalized, interest in Earth's variety into a moral response to the current biological extinction crisis. Biodiversity is both a scientific concept and an ethical stance. To understand this, one must grasp the subtext implied in the word. At first glance, it seems to be just a voguish contraction of the more staid "biological diversity,"[7] which itself is a relatively recent variant of such older terms as "organic diversity" and "natural variety." And indeed, as Rosen and others have freely admitted, "biodiversity" was meant to be a buzzword: one of the goals of the National Forum on BioDiversity (in subsequent usage the software-style spelling went by the boards) was to shove the problem of species loss into the collective face of the U.S. Congress and thereby create a sense of impending disaster. That the Forum did. But, much to the surprise of some of its progenitors, the concept of biodiversity quickly transcended these bare-fisted political origins (Takacs 1996, 37–40).[8] It did so because it has real substance that sets it apart from the usual "latest big thing." Biodiversity is a genuinely compelling and novel concept around which much ecological research and conservation endeavor can be, and is being, reorganized.

What makes it compelling are the overtones of urgency the concept carries. Unlike its predecessors, the word "biodiversity" invokes in scientists and conservationists a penumbra of emotions, a complex mixture of awe and reverence (with respect to the beauty of nature's variety) and dread, sorrow, remorse, and anger (with respect to its impending destruction).[9] They *are* over-tones, not things explicit, and that is the key to the term's acceptability to sci-entists. Biodiversity allows them to express these emotions in a subtle way with-out overtly compromising their ideals of objectivity.[10] By the same token, the tremendous surge in academic research about biodiversity gives scientific credi-bility to the long-standing intuitions of conservationists.

What makes biodiversity novel is the prospect of the world being on the verge of a sixth mass extinction—the first ever at the hands of people. Such peril on such a scale from such a source never even remotely shadowed Darwin's thoughts.[11] The realization that people are threatening to undo the web of life has, it is fair to say, shocked many scientists out of complacency.[12] The emergency atmosphere the National Forum on BioDiversity organizers wished to create was an artifice, inasmuch as it was done purposely for that event and with a specific target in mind, but the worldwide loss of biodiversity is anything but a phony emergency. It is this contemporaneous authenticity, riding a deepening sense of moral urgency, that gives biodiversity its power and distinctiveness.[13]

However, as liberating and refreshing as the biodiversity movement has been to Western field biology, it is far from being so received by others. There are those defenders of the status quo who recognize it as a legitimate threat to entrenched interests and react accordingly. Perhaps more surprising is that, to many indigenous people, the concept of biodiversity is incomprehensible, both spiritually and as a practical matter (Mead 1999, 151). To them, biodiversity is a token of a false dualism between humans and nature, and as such is irreme-diably foreign. Despite its holistic intentions, biodiversity is still stuck in the Western "either/or" disjunctive mode, rather than reflecting indigenous people's "both/and" way of looking at life.[14] In their view, "there is no uni-tary notion of 'nature,' but a set of relations between humans, spirits and clus-ters of species" (Gray 1999, 62).

As with "nature," indigenous peoples also argue that the concept of biological diver-sity is alien because it separates the phenomenon of nonhuman diversity from their knowledge and livelihood. . . . Biological diversity for most indigenous peoples is taken for granted as part of their everyday existence. The point here is not that "nature" or "biodiversity" do not exist, rather that they are part of an approach to the environment which separates what so many indigenous people keep together and simplifies unnecessarily what is so complicated and unique to each people. (Gray 1999, 62)[15]

So what is biodiversity, then? Is the variety of Earth's life forms and processes some kind of ultimate grab bag through which pharmaceutical companies get to rummage for the next Pacific yew, rosy periwinkle, or other plant with medicinal use? Or is it a wellspring of spirituality diametrically opposed to the motives that drive those companies? Or is it a state-owned good that doubles as a symbol of sovereignty in North–South wrangling over global treaties? All of these? None?

Numerous definitions have been offered, and though they vary in the details, they all aim to capture a sense of the sheer broadness of life. The more technical definitions try to do this by reference to the basic hierarchical levels of genes, species, and ecosystems; others simply assign the term a *very* expansive and enveloping meaning. Thus, in the words of the World Conservation Monitoring Centre's (WCMC's) 1992 global biodiversity status report, biodiversity is "commonly used to describe the number, variety and variability of living organisms," and is essentially a synonym for "life on Earth" (Groombridge 1992, xiii). The other global biodiversity assessment to date, published in 1995 by the United Nations Environment Programme (UNEP), defines it almost identically, as "the total variability of life on Earth" (Heywood 1995, 5). David Takacs's recent interviews with American biodiversity scientists elicited many similar responses.[16] Wilson has since aptly noted that "biologists are inclined to agree that it is, in one sense, everything."[17] He then goes on to give the hierarchical definition:

Biodiversity is defined as all hereditarily based variation at all levels of organization, from the genes within a single local population or species, to the species composing all or part of a local community, and finally to the communities themselves that compose the living parts of the multifarious ecosystems of the world. (Wilson 1997, 1)

"The key to the effective analysis of biodiversity," he concludes, "is the precise definition of each level of organization when it is being addressed" (Wilson 1997, 1). This is an important point. The complexity of biodiversity is expressed, first of all, in how the genetic information flow embodied in animals and plants is distributed across the face of the earth, and, secondly, in how these genetic embodiments interact to form ecological communities (which in turn interact among themselves). There are very few uniformities to be found when one compares the distributions and interactions. Accordingly, biologists and ecologists have created numerous measures of diversity to elaborate on the genes–species–ecosystems hierarchy (Magurran 1988). Examples include alpha, beta, and gamma diversity, which measure differences within and among habitats at local and regional levels; taxic and functional diversity at macro-scales; systems diversity, characterized by (among other classifications) biomes, biogeographic and oceanic realms, life zones, and ecoregions; karyotypic variation (which refers to chromosome complements), allelic diversity, heterozygosity, and similar measures at the genetic level; and on and on. The proliferation of analytical approaches is staggering. Literally: Try carrying around UNEP's *Global Biodiversity Assessment* (Heywood 1995)—a truly monumental volume of more than 1,100 pages. But the whole biodiversity enterprise is not just another case of gratuitous academic gigantism. It is fully justified because, as Wilson emphasized, patterns and trends in biodiversity vary greatly depending upon which scale one is analyzing. The biological totality of the world is full of anomalies that shorthand analyses inevitably miss. It is, for example, quite possible for diversity at many disparate localized sites to be on the increase even as the planet at large is in the throes of a mass extinction.[18]

These observations do not vitiate the usefulness of looking at global, long-term trends in biodiversity, to which we next turn. In all that follows, however, we should hold in mind a caveat against carelessly projecting global trends downward to a particular locale without the requisite "ground truthing." Like politics, all biodiversity is local—ultimately, anyway.

A BRIEF HISTORY OF BIOLOGICAL EXTINCTIONS

In the big scheme of things, extinctions are a constant, having occurred in every significantly long period of geological time. Of all the species that ever lived, more than 95 percent (indeed, probably more than 99 percent) are now extinct (Heywood 1995, 202, 208; Sepkoski 1999, 260). Nonetheless, the generative power of evolution has always compensated. Since the emergence of the first nonmicroscopic animals some 600 million years ago, the graph of life generally has been on the rise. Figure 2.1 shows the ups and downs in the number of families of marine organisms as a proxy for all biodiversity.[19] As we move rightward along the graph toward the present, the upward trend is broken only by the five major extinctions in geological history. The depths of the plunges are,

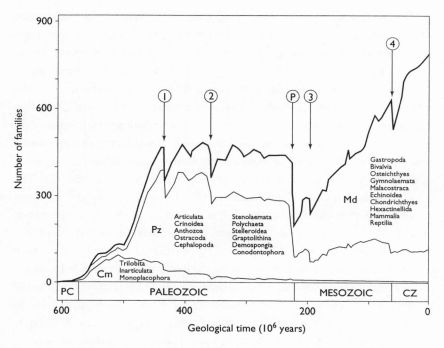

Figure 2.1. The first five major extinction events, as measured by diversity of marine animal families over geological time. The heavy upper curve shows the total number of families with skeletalized species known from the fossil record. The fields below the upper curve show the contributions of the three great evolutionary faunas (Cm = Cambrian; Pz = Paleozoic; Md = Modern). Some of the major classes are listed for each fauna. Arrows mark the five major mass extinctions in the marine realm. P = end-Permian (the largest); 1 = end-Ordovician; 2 = Late Devonian; 3 = end-Triassic; 4 = end-Cretaceous. PC = Precambrian; CZ = Cenozoic. Adapted from Sepkoski (1992), as published in Heywood (1995, 204), from which this caption is taken. Reproduced by permission of Columbia University Press.

with one exception, of similar magnitude, representing losses of roughly 40–80 families each time, or about 10–25 percent of the total number of families. The exception is the third extinction of 245 million years ago, marked by a circled "P" in the graph. This event, known as the end-Permian extinction because it marked the close of the Permian period (and of the Paleozoic era), set biodiversity back almost 300 million years. Fifty percent of the marine families then extant were lost, and as many as 96 percent of the species in them (Heywood

1995, 208). The terrestrial evidence suggests that the end-Permian was a truly comprehensive extinction. It marked a major decline in land plants and insects; it was, in fact, the "only great extinction in the history of insects," with the percentage of familial and species losses exceeding even those in the marine realm. To a degree inadequately known, it also marked a decline in four-limbed terrestrial vertebrates (Heywood 1995, 206–208; quote from p. 206). The "bounce-back time" after the other extinctions was a minimum of several million years, but in the case of the end-Permian extinction it was tens of millions—rather quick by geological standards, though impossibly long by ours.

In Figure 2.2 we see the same five extinction events, this time dramatically marked by thunderbolts, plotted against a timeline stretching back to the beginnings of life. (Each horizontal line is on a different scale.) This graph adds, in the lower right-hand corner, the advent of humankind to the mix. Note how short a time humanoid species have been on the scene—and that there is now a sixth thunderbolt, since we, *Homo sapiens*, have set off the first mass extinction of species in 65 million years. All such extinctions are caused by massive disruptions in ecosystem functioning; some of the hypotheses proposed to account for the first five include climatic change, drops in sea levels, and catastrophic asteroid strikes (Heywood 1995, 209). The current, sixth, extinction is the first where the direct agency of a single species is implicated.[20]

AGRICULTURE AND THE DECLINE OF VARIETY

As measured across millions of years, the world is just starting to descend from the pinnacle of biological diversity. At what point the peak was reached, no one knows exactly, but there is accumulating evidence that taxonomic diversity (both marine and terrestrial) is lower today than it was 5–15 million years ago, probably because of long-term cooling in the global climate (Heywood 1995, 210–211). But if we now shift time scales, leaving the vast stretches of geological time behind and focusing on historical time, that is, the last 10,000 years, we find that the descent is beginning to gather speed.

The beginning of this period saw the dawn of localized agriculture, which arose in several complexes: first in western Asia, in time spreading from there, and later, separately, in China, the Americas, and elsewhere (Nettle 1999a, 103). Although initial small-scale agriculture did not trigger an immediate, catastrophic decline in biological diversity,[21] it set in motion the eventual development of concentrated agriculture, which is responsible for most large-scale dev-

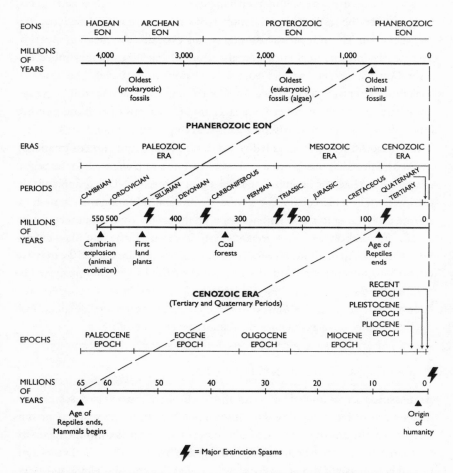

Figure 2.2. Extinction events over evolutionary time. The full geological history of life goes back more than 3.5 billion years, when the first single-cell organisms appeared. Key episodes in evolution are placed within the divisions of geological time: eons divided into eras, eras into periods, and periods into epochs. (The solid time lines are scaled differently.) Extinction events are marked with lightning bolts. The current, sixth extinction is shown at the bottom right. Source: Wilson 1992, 189. Reprinted by permission of the publisher from *The Diversity of Life* by Edward O. Wilson, Cambridge, Mass.: The Belknap Press of Harvard University Press. Copyright © 1992 by Edward O. Wilson.

astation of wild habitat—the leading cause of species endangerment and
ecosystem fragmentation. This occurs either directly, by conversion of lands
and waters to agricultural uses, or indirectly, by supporting livelihoods that
make possible other forms of habitat destruction, such as urban sprawl. It is
true that under certain circumstances early humans did overhunt and over-
gather some species to extinction, and the use of fire by hunter–gatherers to
drive game and clear forests is now known to be far more widespread than previously thought. In particular, there is little doubt that hunter–gatherers had
a tremendous effect on the native biota everywhere during the first period of
colonization.[22] Although there is still debate over the cumulative environmental effects of pre-agricultural societies (because their impacts varied so much
from place to place), it may be stated that, had humans never passed out of the
hunter–gatherer mode, in all likelihood there would be no global extinction
crisis today. The cumulative environmental impact of hunter–gatherers was
large but dispersed, occurring at different places at different times. Their activities did not precipitate extinctions simultaneously on a global scale: there were
too few people and the pressure they brought to bear on the environment was
too spread out.[23]

That dynamic was irrevocably changed by the discovery that some plant and
animal species can be domesticated. The advent of agriculture set off a whole
host of changes in civilization whose cumulative impacts have brought us to
the threshold of the biodiversity crisis; deep-flowing changes, such as allowing
humans to supply their nutritional needs from a much smaller area using fewer
species, and to buffer themselves against seasonal fluctuations in food availability by achieving higher yields. Health and longevity improved, allowing
human populations to grow. Using the most conservative assumptions, during
the period localized agriculture evolved (about 12,000–8,000 B.C.) the average
global population growth rate increased by a factor of ten (Cohen 1995, 78,
400). As time passed, the need for nomadism declined, and concentrated,
sedentary societies formed.[24] These eventually set the stage (during 1400–1800
A.D.) for the rise of European exploration and out-migration,[25] and, in the 18th
century, the beginnings of urban industrial society and the rise of global-scale,
intensive agriculture, in which the cultivars developed on different continents
were pooled. All this coincided with another surge in population growth rates
(Cohen 1995, 29–30).

When vast areas began to be committed to monoculture row crops or wood-
fiber production or the grazing of a handful of livestock breeds, and when pro-

duction decisions started to be driven by regional or global markets rather than local needs, the devastation of natural variety swung into high gear. Intensive farming, livestock raising, and pastoralism simplify the environment, and are now so widespread that they are a major factor in extinctions through direct conversion of wildlife habitat. It is this kind of agriculture that has allowed humans to arrogate to themselves at least 40 percent of Earth's net primary products of photosynthesis (Vitousek et al. 1986)—a startling figure, and, to my mind, the single most telling environmental statistic yet produced. As Niles Eldredge has observed, this is what "brought *Homo sapiens* into the extinction business in a truly global and all-encompassing way," transforming "the very stance that our species holds vis à vis nature" (Eldredge 1995, 81–82). Up until the last three centuries or so, humans interacted with local ecosystems primarily at a local level:

> The world before the industrial revolution can be envisioned as a mosaic of co-evolving social and ecological systems. Within each area of the mosaic, species were selected for characteristics according to how well they fit the evolving values, knowledge, social organization, and technologies of the local people. At the same time, each of these components of the social system was also evolving under the selective pressure of how well it fit the evolving ecological system and the other social components. Local knowledge, embedded in myths and traditions, was correct, for it had proven fit and through selective evolutionary pressure, had become consistent with the components of social and ecological systems it explained. (Norgaard 1988, 206–207)

Both cultural facets and species spilled over areal boundaries within the biocultural mosaic, leaving a "pattern of coevolving species, myths, organization, and technology" that was "patchy and constantly changing" (Norgaard 1988, 207). Now it has become "more important to interact with other people living elsewhere than with other species occurring locally" (Eldredge 1995, 82). Agriculture, we might say, has changed from an endeavor where production is dispersed and distribution concentrated to one where production is concentrated and distribution dispersed.

However, even today many forms of agriculture are not inimical to biodiversity. Traditional agroecosystem techniques (for example, intercropping and shade cropping), especially when practiced by small groups of people scattered across a landscape with large areas of interspersed wild lands, probably have a negligible effect on biological diversity—and not infrequently enhance it, at least on a local scale (Plenderleith 1999, 287–291). Before we judge the impacts of agriculture, we need to distinguish among low-intensity agriculture aimed at subsistence needs and local markets, modern export-oriented agribusiness, and the gradations in between. "Some recently 'discovered' 'cultural land-scapes'," cautioned the anthropologist Darrell Posey, "include those of ab-original peoples who, 100,000 years before the term 'sustainable development' was coined, were trading seeds, dividing tubers and propagating domesticated and non-domesticated plant species . . . to the extent that it is often impossible to separate nature from culture" (Posey 1999b, 7).

EXTINCTION RATES: MEASURING THE PAST, PREDICTING THE FUTURE

The relative impacts of agriculture, industrialization, and population growth on species extinctions over the last 10,000 years are a fascinating matter of debate. The argument so far is inconclusive, but clearly we have entered a new phase in terms of how quickly species are disappearing. Among the majority of biologists, there is no dispute that humans have begun to set off another mass extinction (Cincotta and Engelman 2000, 14).[26] These scientists will readily admit that they can't put solid numbers to this. But, however the extinc-tion rate is reckoned, the general conclusion is the same: The current rate is far above previous levels—perhaps 1,000 times higher—and the cause is human activity (Pimm and Brooks 2000, 59).

What is the basis for this conclusion? An important question, since it is the cornerstone of the claim that we are in a biological extinction crisis. In an empirically based review, May, Lawton, and Stork (1995) have clearly laid out the steps that must be taken to arrive at a defensible estimate of the current extinction rate. They also properly emphasize the limitations of the data used upon which each step of the process depends.

First, one needs a benchmark. The only way to determine whether humans are killing off species faster than they would die out otherwise is to estimate the background extinction rate over time. To get it, one must know the aver-age life span of a species, from origination to extinction. This is determined by examining fossils for a wide range of organisms. Allowing for the fossil record's numerous imperfections, it is thought that the average species has a life span of 5–10 million years (May, Lawton, and Stork 1995, 2–3), though recent evi-dence of shorter spans for some species suggests more variability, perhaps 1–10 million years (May 2000, 41–42; Sepkoski 1999, 261).[27]

Next, one must estimate the total number of species that ever existed. One way to do this is to project backward from the current estimated number of species. If we adopt a figure of 5–10 million for that number (which is conservative), apply an average life span of 5–10 million years to each species, and note that the fossil record covers about 600 million years, then it follows that the current number of species represents about 1–2 percent of all those that have ever lived. But one must adjust for the fact that most species are terrestrial, and weight the percentages accordingly. Thus the current number of species actually represents about 2–4 percent of all those that have ever lived. To restate this negatively, about 96–98 percent of all species that ever existed are now extinct.[28]

Over the 600-million-year period, this averages out to a background extinction rate of one species per year—though because of the vagaries of the fossil record, it could easily be twice as large or half as small (May, Lawton, and Stork 1995, 6). Using somewhat different assumptions, the UNEP biodiversity assessment gives a background rate of two to three per year, and the WCMC assessment has it at four per year (Heywood 1995, 202, 232; Groombridge 1992, 197).[29] But, as the WCMC points out, "even if background rates were ten times higher than this, extinctions amongst the 4,000 or so living mammals would be expected to occur at a rate of around one every 400 years, and amongst birds at one every 200 years. It is indisputable that the extinction rate in recent times has been far higher than this and that man has been the overwhelming cause" (Groombridge 1992, 197).

This brings us to the question of how to estimate current and future rates of extinction. By far the most prevalent approach is to take the species-area relationship, a known empirical rule derived from island biogeography (MacArthur and Wilson 1967), and combine it with estimated rates of habitat loss (e.g., of tropical deforestation). The rule can be expressed as follows: if 90 percent of an area of habitat is lost, 50 percent of the species found in that area will disappear. There are other ways to estimate current and future rates,[30] but the key point is that they each "give roughly concordant answers," and by any measure these rates are far higher than the background (Heywood 1995, 232; May, Lawton, and Stork 1995, 18).

As we have seen, over a long span of time and in the absence of undue human interventions, one would expect something on the order of one to four species to become extinct every year. During the past four centuries, 1,138 species are recorded by the WCMC as having become extinct—"extinct" mean-

Table 2.1

Species Extinctions since 1600

Species	WCMC[a] 1992 Global Biodiversity Status Report		IUCN[b] 1997 Red List of Threatened Plants	
	Total Described Species	Recorded as Extinct (% of Total)	Total Described Species	Recorded or Suspected as Extinct (% of Total)
Molluscs	100,000	191 (<1)	—	—
Birds	9,500	115 (1.2)	—	—
Mammals	4,500	58 (1.3)	—	—
Other animals	~1,300,000	120 (<1)	—	—
Total, animals	~1,400,000	484 (<1)		
Fern allies	—	—	1,318	4 (<1)
True ferns	—	—	9,053	18 (<1)
Gnetophytes	—	—	40	0 (0)
Conifers	—	—	586	2 (<1)
Gingko	—	—	1	0 (0)
Cycads	—	—	180	3 (1.6)
Dicotyledons	—	—	167,224	556 (<1)
Monocotyledons	—	—	63,611	168 (<1)
Total, vascular plants	240,000–270,000	654 (<1)	242,013	751 (<1)
Total, all higher plants and animals	~ 1,700,000	1,138 (<1)	—	—

[a] World Conservation Monitoring Centre data: Groombridge 1992, 200 (repeated in Heywood 1995, 233).

[b] International Union for the Conservation of Nature data: Walter and Gillett 1998, xxxv–xliv. Figures from Walter and Gillett include species not definitely located in the wild during the past 50 years (IUCN Red List category "Ex"; 380 species of the above total) as well as those that are suspected to have recently become extinct (category "Ex/E"; 371 additional species).

ing that they are no longer known to exist in the wild after repeated searches of their type localities and other known or likely habitats (Table 2.1). This, an average loss of about 2.8 species per year, seems to fall right in with the long-term background rate. But, as researchers familiar with the topic would be the first to say, the numbers shown in Table 2.1 grossly undercount the extinctions over the period. Birds and mammals are disproportionately well studied among organisms, so their extinction numbers are reasonably accurate. Those of the other, much more numerous categories are not: "other animals" includes insects, for which only 61 extinctions have been recorded—33 of which are of highly conspicuous butterflies, many of them from Hawaii. This impossibly small number "speaks eloquently of where the researchers, rather than the endangered species, live" (May, Lawton, and Stork 1995, 13). The paucity of taxonomists working in the tropics, where the great majority of species are, also ensures that the figures in Table 2.1 are an undercount. In sum, the record of recent extinctions is both radically incomplete and skewed heavily toward the birds, mammals, butterflies, flowering plants, and similar conspicuous species that are found in temperate and Mediterranean climates where the majority of scientists live. These groups account for a disproportionately high share of the recorded extinctions, even though they constitute only a tiny fraction of all species. Certainly the true number of species extinctions over the last four centuries is much higher than the recorded tally.

Another telling way to look at the data of Table 2.1 is to recognize that about 1 percent of birds and 1 percent of mammals have been recorded as becoming extinct since 1600.[31] If the same proportion holds for insects (of which there are at least 1 million and probably more like 3 million species), then the number of extinctions among them over the period 1600–1995 goes from 61 to 10,000–30,000.[32] This alone would raise the recent overall extinction rate from 2.8 to 25–75 species per year. Because even less is known about plant extinctions, it is impossible to say whether the very low proportion of recorded plant extinctions is an accurate figure or an artifact of poor data. The current (1997) *IUCN Red List of Threatened Plants*, the most comprehensive assessment ever done, reports only slightly more extinctions (751) than does the WCMC (Table 2.1). "Given the often quoted predictions in the popular and scientific press of extremely high extinction rates, the numbers in this book will seem very low," the *Red List* editors remark.

> While this may in part be due to a lack of data and the fact that plant extinctions have only recently been recorded, at the same time it may be that we are still at the brink of a far more serious wave of extinction. The grim statistic that at least 6,522 species are [listed as] Endangered, and that these are likely to join the Extinct list in the near future, should be taken as a warning that the situation will get far worse unless vastly increased conservation action is taken now. (Walter and Gillett 1998, xlvi)[33]

Whatever the true rate over the past 400 years, most biologists agree that the rate over the next century will be far higher. All sorts of projections have been published in the last twenty years (Table 2.2). Most of them are based on species-area relationships and are predicated upon a continuance of current rates of deforestation in the species-rich tropics. The earliest (ca. 1980–1985) estimates, projecting a 20–50 percent overall species loss by 2000, were too high (Groombridge 1992, 202–203).[34] More sophisticated models of extinction, and better data on deforestation, have lowered subsequent estimates. Roughly speaking, more recent studies predict a 1 percent to a 9 percent global loss per decade over the next 25–30 years. Wilson says that if the present rate of environmental destruction continues, there is a strong possibility of a 20 percent loss of global biodiversity over the next 30 years—a 6–7 percent loss each decade (Wilson 1992, 278–280). In a summary analysis based on recent theoretical and empirical work on biodiversity "hot spots" and tropical habitat fragmentation, Stuart L. Pimm and Thomas M. Brooks conclude that between one-third and two-thirds of all species will be lost if current trends continue (2000, 59).

It is important to understand that most of these are estimates of the number of species that will be *committed to extinction* through habitat loss during the period in question, not the number that will actually become extinct during that time. There is always some delay between the time a species falls below its minimum viable population size and the death of its final individual. For longer-lived species, the delay can be considerable. Species that are beyond hope of recovery have been referred to as "the living dead."

To get an idea of what the percentages just cited mean in terms of species lost per year, per day, and per hour, Wilson asks us to consider this: if one looks at the tropical rain forests alone, choosing extremely cautious parameters so as to arrive at the most optimistic possible scenario, at the current rate of destruction 27,000 species are committed to extinction each year. That is 74 per day, or 3 every hour (Wilson 1992, 278–280).

This, like all predictions of biodiversity loss, is a guess: a well-educated guess, but a guess all the same. There are two reasons why we can do no better than guess. First, because the immediate causes of species decline and extinction are highly variable from place to place. Second, because scientists do not know, are not even close to knowing with any exactitude, how many species exist on Earth. More than 1.7 million have been formally described in the literature, and thereby given the Linnaean binomial tags that serve to locate them within the collective scientific mind of the West.[35] A conservative estimate of the total number of species is 5 million, with 5–15 million seeming a probable range (Stork 1997, 65), but some biologists think the real number is far higher—per-

Table 2.2

Estimates of Potential Species Extinctions

Estimate	Average Loss per Decade (%)	Basis of Estimate/ Assumptions	Reference
1 million species, 1975–2000	4	Extrapolation of past exponentially increasing trend	Myers 1979
15–20% of species, 1980–2000	8–11	Estimated species-area curve; forest loss based on Global 2000 projections	Lovejoy 1980
1 million species or more by 2000	4	If present land-use trends continue	National Research Council 1980
As much as 20% of all species	—	Unknown	Lovejoy 1981
50% of species by 2000	25	Different assumptions and an exponential function	Ehrlich and Ehrlich 1981
Several hundred thousand species in a few decades	—	—	Myers 1982
25–30% of all species, or from 500,000 to several million by 2000	—	—	Myers 1983
500,000–600,000 species by 2000	2	—	Oldfield 1984
750,000 species by 2000	3	All tropical forests will disappear and half their species become extinct	Raven 1986
20–25% of all species, 1986–2011	10–12	Present trends will continue	Norton 1986
15% of all plant species and 2% of all plant families by 2000	—	Forest loss will proceed as predicted until 2000 and then stop completely	Simberloff 1986
12% of plant species in neotropics; 15% of bird species in Amazon Basin	—	Species-area curve ($z = 0.25$)	Simberloff 1986
2,000 plant species per year in tropics and subtropics	8	Loss of half the species in area likely to be deforested by 2015	Raven 1987
9% of species by 2000	7–8	Based on Lovejoy's calculations using Lanly's (1982) estimates of forest loss	Lugo 1988
25% of species, 1985–2015	9	Loss of half the species in area likely to be deforested by 2015	Raven 1988
At least 7% of all plant species	7	Half of species lost over next decade in 10 "hot spots" covering 3.5% of forest area	Myers 1988
0.2–0.3% per year	2–3	Half of species assumed lost in tropical rainforests are local endemics that become extinct with forest loss	Wilson 1988, 1989

Table 2.2 *continued*

Estimate	Average Loss per Decade (%)	Basis of Estimate/ Assumptions	Reference
5–15% of forest species by 2020	2–5	Species-area curve ($0.15 < z < 0.35$); forest loss assumed twice rate projected by FAO for 1980–1985	Reid and Miller 1989
2–13% loss, 1990–2015	1–5	Species–area curve ($0.15 < z < 0.35$); range includes current rate of forest loss and 50% increase	Reid 1992
27,000 species per year	—	Species lost by reduction of tropical area only ($z = 0.15$)	Wilson 1992
20% of all species by 2022	6–7	If present rate of environmental destruction continues	Wilson 1992
Red Data books for selected taxa: 50% extinct in 50–100 years (palms) and 300–400 years (birds, mammals)	1–10	Extrapolating current recorded extinction rates and using the dynamics of threatened categories	Smith et al. 1993a, 1993b
1,100–2,000 extinctions per million species per year	—	Extinction probabilities from vertebrate Red List categories	Mace 1994
10,000 extinctions per million species per year	—	Increasing human energy consumption	Ehrlich 1994
100,000–500,000 insect species in next 300 years (assumes 8 million insect species total)	0.0004– 0.02	As in Smith et al. 1993a, 1993b, as applied to insects	Stork 1997 (citing Mawdsley and Stork 1995)
1,000 extinctions per million species per year	—	Half-life of 50 years for threatened birds	Pimm and Brooks 2000

Sources: Table derived from Lugo 1988, 59; Groombridge 1992, 203; Wilson 1992, 278–280; Stork 1997, 60–63; Pimm and Brooks 2000, 59.

haps 30 million or more.[36] And even for scientifically described species, "accurate information on status and abundance is available for only a tiny proportion" (Groombridge 1992, 198). All in all, knowledge of biological extinction past, present, and future is still scant.

It is also misleading to focus exclusively on species extinctions as losses in and of themselves. The extirpation of discrete populations, even those of species not threatened with full extinction, is a considerable diminution of biodiversity. Yet no one is keeping track of population extinctions (Ehrlich 1995, 218). Even harder to measure, yet vital to the functioning of ecosystems, are the interactions within and between species (Janzen 1988a, 132). This is an area of special interest to scholars of coevolution (Thompson 1994). The extinction of a species occupying a key position in a network of relationships (e.g., mutu-alisms) can set off a series of changes (including other extinctions) whose net result is the loss of interactions that held the ecosystem together. It is possible, therefore, for the loss of single species to disrupt an entire ecosystem.

It has to be conceded that projections of biological extinctions involve a leap of faith from a shaky empirical platform. Little can be known with certainty about past extinction rates, yet they are the basis for all inferences about the comparative severity of the current wave of extinctions. Skeptics will always have this as a weapon. Yet the fact that most biologists (who are themselves trained to be skeptical of broad claims) are in consensus is impressive. Their view is worth adopting as a working hypothesis until credible and convincing counter-theories are presented.

To sum up: The major trend in biodiversity is a widespread, radical decline, measured at any level of biological organization. Many subsidiary trends, con-sequent upon this, are being played out at local and regional scales. There are also other global trends, such as global warming and population growth, that intertwine (Cincotta and Engelman 2000). All of them are the result of people's actions. It is impossible to escape the human dimensions of biodiversity. The use of the plural "dimensions" is intentional, for the impact of humans on bio-diversity (and vice versa) is not monolithic. To fully understand what is hap-pening to biodiversity, we must also understand what is happening to cultural diversity—the second great realm of living difference.

WHAT IS CULTURAL DIVERSITY?

"Cultural diversity" may be defined as the variety of human expression and organization, including that of interactions among groups of people and between these groups and the environment. Thus defined, the term is global in scope, both physically and conceptually. It comprises the entire index of dis-crete behaviors, ideas, and artifacts as exhibited by the whole of humankind, no matter where they live. It also covers the complete range of types of cul-tural interaction, as well as their outcomes or manifestations.

At the outset, we must clearly distinguish this global meaning of cultural diversity from that which the term has acquired in some arenas of political dis-course. In the United States, "cultural diversity" has become a cover term refer-ring to the redress of racial, ethnic, and gender inequities in society. It sags under the weight of onerous, highly politicized associations. I say "associations" rather than "meanings" because much of what is said and written in this American context is intended as a tendentious shorthand, is more or less intentionally vague, is not intended to carry forward anything productive, and will never terminate in a resolution. The resulting trauma has been well described by the religious scholar Martin E. Marty in his essay on pluralism and unity in the United States, *The One and the Many*. In brief:

> During the final quarter of the twentieth century many groups of citizens have come to accuse others of having wounded them by attempting to impose a single national identity and culture on all. The other set, in turn, has accused its newly militant adversaries of tearing the republic apart. They do this, it is said, by insisting on their separate identities and by promoting their own mutually exclusive subcultures at the expense of the common weal. Taken together, these contrasting emotions produce a shock to the civil body, and a paralysis of the neural web of social interactions. (Marty 1997, 3)

One of the casualties of this "almost shattering controversy" has been the legitimate vocabulary of cultural diversity. The very word "culture" has become loaded, as in the phrase "culture wars," a favorite of the magazines. "Diversity" is even more abused, pumped so full of hot air that it has become nearly meaningless. For example, there is now the self-described profession of "diversity consultant": a specialist who advises companies on how to avoid sexual harassment lawsuits. Obviously, "diversity" is here meant to carry connotations of enlightened respect and sensitivity toward others—a laudable aim, to be sure, but not a very precise usage. Then we come to a word such as "multiculturalism." This ought to be a simple description of the condition that necessarily prevails in a pluralistic society. Instead, it has been turned, cynically and with calculation, into a term of opprobrium used to caricature a supposed political agenda.[37] It would be convenient, when discussing global cultural diversity, to refer to "multiculturalism" as a positive response to the challenges of pluralism, as a constructive strategy to encourage acceptance of cultural difference. But for American readers, at least, the word has been made over into little more than a sneer.

Marty takes care of the problem by categorically excluding such debased terms from his narrative. A practical solution, but "cultural diversity" is too apt and too important to abandon. That is why its global meaning must be carefully marked off; in the sense I wish to use the term, it has no prefixed political connotations. It encompasses all aspects of culture, regardless of their social status or political posture within a particular society. "High" culture, "bourgeois" or "middlebrow" culture, "folk" culture, "mainstream" culture, "underground" culture, "counter-culture"—all such sociological terms are included, though their analysis is not our objective. Rather (to restate the global definition one final time) we are taking "culture" at its broadest, so that it is defined as any and all products of the human intelligence, with "cultural diversity" meaning the range of difference to be found among those products.

GLOBALIZATION AND CULTURAL DIVERSITY

As with biodiversity, the major trend in cultural diversity is rapid diminishment. The idea that culture is becoming less varied is not new. In the 1850s the French positivist philosopher Auguste Comte set out with remarkable prescience the future of globalization. He predicted that five key characteristics, already then present in European society, would eventually spread throughout the world: industrial life and the organization of labor, the homogenization of aesthetic tastes, international agreement on the methods and contents of science, a preference for the democratic republic as the political form, and morality based not on theology but on humanism. A century and a half later, we see that Comte was very close to the mark, as far as these generalities go. The rest of his philosophical apparatus is less impressive. He was much given to spinning out overelaborate "phases" of history that humanity would pass through as we move, inevitably, toward the single positivist society that he called the "ultimate regime." No detail of the coming utopia was too small for Comte to specify, to the point where the valuable aspects of his thought became lost amid a zealous quirkiness (Todorov 1993, 27–30).

But if Comte was a crank, he was a crank who got his general trends right. Today plenty of people share the outlines of his vision, if not the overwrought details. Where Comte was wrong, spectacularly wrong, was in his insistence that everyone would naturally fall in, single-file, with a global unification program. It hasn't happened so far. There are of course the dissidents one might expect on the far ends of the political spectrum, yet there is also a growing body of dissent that cannot be so easily labeled. For example, one can interpret the fast rise of the biodiversity movement, which values all species regardless of their usefulness to humans, as a tacit dissent against the dominant materialistic social and economic dogmas of our day. A similar against-the-grain attitude has infused at least some quarters of every social science for a much longer time—most notably (for our purposes) anthropology and linguistics, the two disciplines most consistently interested in cultural diversity per se.

As we saw in the last chapter, the sources of globalization were put in place during the formative period of Western civilization. The difference now is that globalization runs in overdrive, fired by one advance after another in telecommunications and information processing, orchestrated by transnational corporations determined to develop a seamless global marketplace. National governments have generally converged with corporate interests on a so-called free market ideology that values uniformity and consistency (to better deliver products and profits) and devalues variety and uncertainty (as barriers to trade and investment).

The last quarter-century has seen dozens of political critiques of globalization (e.g., Schiller 1976; Smith 1980; Boyd-Barrett 1982; Mattelart 1982; Hamelink 1990; Tenbruck 1991; Shiva 1993; Barber 1995; Mander and Goldsmith 1996; Skutnabb-Kangas 2000; Toledo 2001), and, given the recent protests against the World Trade Organization, The World Bank, and the International Monetary Fund, many more are no doubt on the way.[38] It is not my aim in this book to produce another one. Nevertheless, it is important to grasp the outlines (if nothing else) of the political argument against globalization, particularly in relation to cultural diversity.

In this view, there is a more or less conscious attempt by Western interests to impose the "perspective of global monoculture" on the rest of the world:

> This vision offers a universal and only slightly varying set of activities and expectations for the entire planet, a homogenized directory of standards for everything from diet and clothes to transportation and architecture. Global monoculture dictates English lawns in the desert, business suits in Indonesia, orange juice in Siberia, and hamburgers in New Delhi. It overwhelms local cultures and "develops" them regardless of the effects on cultural coherency or capacities of local ecosystems (Berg 1981).

Or, as Jeffrey McNeely puts it, "consumers in Bangkok, Bogotá, Bangui, Boston, Brisbane, and Belfast are eating the same Big Macs, drinking the same Pepsi, watching the same Bill Cosby programme, smoking the same Marlboros, sending faxes over the same Toshiba modem, and wearing the same Levi's jeans" (McNeely 1995, 1). In a recent radio talk, the economist Lester Thurow described how, when traveling in Saudi Arabia, he happened upon a

Bedouin encampment in the Empty Quarter. Everyone was clustered around the TV hooked up to a portable satellite dish, running off of a Honda generator. Only a few years ago such stories sounded shocking; now they are commonplace.[39] We seem bent on reducing civilization to a single type, that of Western (largely U.S.) post-industrial society—a kind of Global America. And, to hear the average CEO tell it, the move is not just inevitable, it's inevitably good. Comte would be welcome in many a boardroom today.

Leaders of ethnic and indigenous groups, academicians, labor and human rights activists, and others have decried what they see as the steamrollering of cultural variety and autonomy (Slikkerveer 1999). The emergence of a global market economy, built upon the unprecedented power and range of transnational corporations, is often given as a reason (Mattelart 1982). Other critics cite colonialism and its successors (e.g., "American cultural imperialism"; see Schiller 1976; Smith 1980); the increasing reach of telecommunications, which is chiefly controlled by developed countries (Boyd-Barrett 1982; Hamelink 1990) and which saps all cultures of their autonomy as they are drawn into an electronic free-for-all (Tenbruck 1991); and the opening up of remote areas with roads and airstrips.

Whether cultural diversity is truly on the wane is a matter of vigorous debate. There is a large and divergent literature pertinent to the question in the fields of ethnology, cultural anthropology, sociology, linguistics, communications theory, cultural studies, and other social sciences. Much of this writing further concerns whether cultural diversity is good or bad. To indicate the range of the argument, some theorists think world harmony can arise only from a single global state using one universal language,[40] whereas others see the answer in the break-up of today's state-dominated polity into thousands of small self-governing entities based on local cultures (Wittbecker 1991). And, it should be noted, for some theorists the notion of cultural diversity is a misleading simplification (Featherstone 1991).

THE VALUE OF CULTURAL DIVERSITY

Be that as it may, concern about the loss of global cultural diversity is on the rise. As with biodiversity, cultural diversity is emerging as a central organizing concept for scholarly research and activism. Why so? Why is cultural diversity important? The central argument I am developing in this book is that it is a crucial part of our common humanity at its deepest level. But there are other reasons, a few examples of which may be briefly noted here.

Cultural diversity is a reservoir of creativity, "fostering greater esthetic, intellectual, and emotional capacities for humanity as a whole" so that we arrive "at a higher stage of human functioning" (Fishman 1982, 6–7). This creativity is not confined to the arts; it is also a source of potential solutions to social and environmental problems, solutions that would otherwise be ignored by politically dominant cultures precisely because dominance breeds complacency and stunts the capacity for self-criticism. In this sense, cultural diversity is an indispensable corrective or counter-balance.

This suggests a further value to maintaining cultural diversity: it is a matter of social justice. The reason certain cultures achieve political dominance is, at least in part, because they have subjugated other cultures against their will. It will suffice to simply mention two major trends in recent history, European-based colonialism and the spread of Marxism, both of which are replete with examples. Against this backdrop, recognizing the right of people to choose their own languages, worldviews, and lifestyles becomes a way to redress past injustices (cf. Thieberger 1990, 348–350).

Apart from its value to the world's peoples as a whole, the separateness of cultures is critical to the well-being of individuals who identify themselves with one in particular. For example, many indigenous people are certain that speaking their native language is a crucial part of who they are. More broadly, cultural identification is the basis of ethnicity, and the central value of ethnicity is that it imparts an affirming sense of uniqueness to individuals while involving them in a group in a way that is consonant with our biological imperatives as social animals.

On a purely intellectual level, the components of cultural diversity are the raw materials of scholarship in anthropology, sociology, history, literature, musicology—of every field in the social sciences and humanities. When cultural diversity is diminished, the value of future research in all these fields is lessened too. This is so even in subjects such as linguistics, which has been dominated over the past generation by theoretical research, especially the search for universals drawn from a small sample of languages. As we will shortly see, an increasing number of linguists are not satisfied to let their discipline "go down in history as the only science that presided obliviously over the disappearance of 90% of the very field to which it is dedicated" (Krauss 1992, 9–10; see also Corbett 2001).

Finally, it should be noted that all these values of cultural diversity deliver economic benefits to the world's peoples, either directly as income (e.g., through expenditures on the arts) or indirectly as avoidance of costs (e.g., by offering novel solutions to social problems that require outlays of money to remedy).

Of course, the benefits of cultural diversity are disputed by those who take Voltaire's solution literally—who are convinced that social cohesion comes only through uniformity. Even proponents of cultural diversity must acknowledge that the benefits are bound up with risks, risks that are all too evident in the seemingly never-ending rounds of ethnic conflict that are a staple of the news.

INDICATORS OF CULTURAL DIVERSITY

Nonetheless, if we want to know which way cultural diversity is going (even if only in a general sense), we need some signposts. Indicators of cultural diversity can be divided into three classes (Table 2.3). The first is directly related to subsistence and livelihood: it includes the selection of crops and diet, land management strategies, environmental and medical knowledge, and the use of language. The second comprises creative activities: architecture, literature, music, oral history and story telling, and the visual and performing arts. The third is defined by a complex social process of identification with, and acceptance by, a self-distinguished group. It includes ethnicity, religion (and, increasingly, its modern counterpart, secular affiliations), and social organization.[41] All three classes overlap to some degree.

If one could analyze each of these indicators and relate the results on a global scale, clearly the overall thrust would be that cultural diversity is declining. The trends in some indicators are unequivocal. Selection of crops and choice of diet are good examples. Domestication of wild plant species is a cornerstone of agriculture and a hallmark of many cultures. Traditionally, farmers cultivated genetically distinct varieties to be safe in case one failed. Raising crops for cash and export, the rise of agribusiness, and the "Green Revolution" have all beckoned farmers to the same genetically similar varieties. The spectrum of agricultural (as well as forest and fisheries) products is getting steadily narrower. Some 1,500 local rice varieties have recently disappeared from Indonesia because four out of every five farmers stopped planting them in favor of a few new strains [World Resources Institute (WRI), IUCN, and UNEP 1992, 9].[42] In the Terai region of Nepal, only one rice variety is now popularly cultivated, though some 1,800 local varieties have been recorded (Upreti 2000, 329, citing Shrestha 1999, 33). Not only does this undermine biodiversity,

Table 2.3
Some Indicators of Cultural Diversity

Subsistence and Livelihood	Creative Activities	Group Identification
Crop selection, animal breeding and husbandry techniques, land management strategies	Architecture	Ethnicity
	Dress and personal adornment	Religion
	Literature	Secular affiliations
Diet	Music	Social organization and
Environmental knowledge	Performing arts	kinship systems
Language	Storytelling and oral history	
Medicine and medical practices	Visual arts	
Recreation		

Source: Adapted from Harmon 1998a, 354.

it tosses away thousands of unique achievements of the human intellect, each finely tuned to a local environment and painstakingly arrived at.

The shift to a narrow crop genetic base implies that the diets of the world are becoming more similar. Verifying this would require an extensive survey of food and drink in developing countries along the lines of the Organisation for Economic Co-operation and Development (OECD) series on Europe and North America. Trends in at least one commodity, however, strongly suggest what might be happening. Refined sugar is a key ingredient in many of the highly processed foods and drinks that are a staple in Western diets. Some forty years ago, annual consumption in developing countries was low—less than half that in developed countries. By the early 1980s it had outstripped that in the developed world, and now is nearly twice as much (Figure 2.3).

Among tribal peoples, traditional food habits are remarkably resistant to voluntary change, except for certain instances where imported foods attain prestige because they become identified with powerful outsiders. More often,

Dietary changes are forced upon unwilling people by circumstances beyond their control. . . In some areas, new foods have been introduced by government decree, or as a consequence of forced relocation or other policies designed to end hunting, pastoralism, or shifting cultivation. Food habits have also been modified by massive disruption of the natural environment by outsiders—as when sheepherders transformed the Australian Aborigines' foraging territory, or when European invaders destroyed the bison herds that were the primary element in the [North American] Plains Indians' subsistence patterns. Perhaps the most frequent cause of

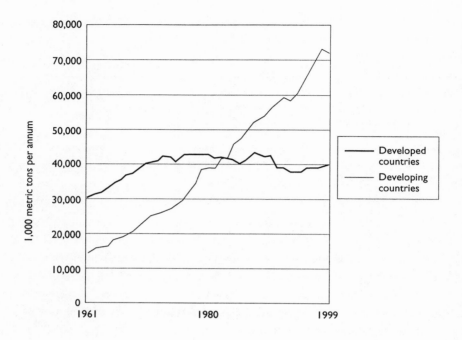

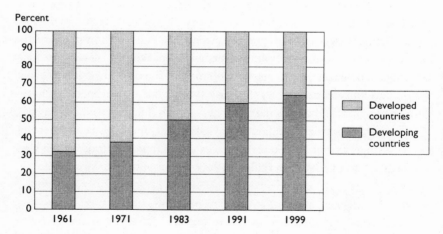

Figure 2.3. Annual use of refined sugar for food, 1961–1999. Top: Annual use in developing countries surpassed that in developed countries in 1983. Bottom: In 1961, developing countries accounted for 32% of the annual use; by 1999, their share had doubled. Source: Food and Agriculture Organization of the United Nations, FAOSTAT database, Food Balance Sheet Report Form: http://www.apps.fao.org>nutrition.

diet change occurs when formerly self-sufficient peoples find that wage labor, cash cropping, and other economic development activities that feed tribal resources into the world market economy must inevitably divert time and energies away from the production of subsistence foods. Many developing peoples suddenly discover that, like it or not, they are unable to secure traditional foods and must spend newly acquired cash on costly, and often nutritionally inferior, manufactured foods. (Bodley 1982, 155–156)

The result has been a dramatic rise in malnutrition, dental disease, and nutritional-deficiency disorders.

Artistic indicators of cultural diversity are harder to analyze. It is possible, for example, to compare the state of certain artistic activities from country to country. There are statistics on publishing, libraries, cinemas and cinematic performances, performing arts events, and museums. Many people are alarmed at the recent spate of mergers in which entertainment conglomerates have swallowed up publishing houses, fearing that their books will become more trivial. Others are concerned about the "corporatization" of symphony orchestras, ballet companies, museum exhibits, and so forth. But is there any meaningful way to measure the effects of such changes? And how does one assess the balance between innovation in the arts and the erosion of older styles and forms? For example, is "world music" a synergistic and enriching blend of styles, or a commercial pastiche crowding out authentic, traditional folk music? These are questions with no simple answers (and many people contend that it is absurd to even ask). In any event, international comparisons of artistic data, where such are available, have to be approached with caution:

> In older, more prosperous countries, where the physical necessities of life are in secure supply, more money is available for cultural activities—and, indeed, for collecting data on them—than in less developed countries. Yet a developing country with an embryonic statistical system may have a flourishing cultural life that includes theatrical performances, live music, or the practice of arts no longer central to the Western experience, such as oral storytelling, ceremonial dance, traditional community ritual, or puppetry. Such activities may be more fully integrated into the life of the people than the more measurable cultural pursuits of a developed society. (*Encyclopædia Britannica* 1990, 789)

Important as they are, artistic endeavors are too disparate to serve as the basis for a comprehensive assessment of the world's cultural diversity.

It is evident that we need to look instead to those indicators with the potential for universal applicability. Ethnicity is one possibility. No enumeration of the world's peoples can claim to be authoritative, but the most detailed I have seen is the ethnolinguistic classification given in the second edition of the *World Christian Encyclopedia* (Barrett, Kurian, and Johnson 2001). It gives data on eleven levels, including one comprising 12,583 "constituent peoples and sub-peoples" that are referenced to 238 states and other political divisions. Based on this classification, the authors also published a formula for measuring "cultural distance," that is, how different two cultures are from each other (Barrett, Kurian, and Johnson 2001, 2:16–18; see also Barrett 1982, 107–115).[43]

Table 2.4 summarizes the classification at the level of "ethnolinguistic family," of which 71 are identified, and presents population trends for each family for the period 1900–1980.[44] Considered at a global level, the figures suggest a broadening in ethnolinguistic diversity over the period. In 1900, the five most populous ethnolinguistic families—Chinese, North Indian, Germanic, Slav, and Latin—accounted for 68 percent of the world's people. By 1980, that figure had fallen to 55 percent, with four of the five lagging behind the world growth average. Latin and Slav populations, in particular, grew much more slowly than the world average. Meanwhile, small ethnolinguistic families, such as Spanish- and Portuguese-speaking Afro-American, Negrito, and Mestiço, grew at more than three times the world average. Several larger ethnolinguistic families grew at least one and a half times as much: Iranian, Blanco, Mestizo, Arab, Indo-Malay, Korean, Tai, Viet-Muong, Bantu, and Guinean.

When the world's ethnic groups are looked at on a finer scale, however, there is evidence of a decrease in overall diversity because of the extinction, some of it genocidal, of tribal cultures. Catastrophic drops in tribal populations after contact with Europeans are recorded throughout the anthropological and historical literature. These are not limited to the distant past. To take just one example, it is estimated that 87 of the 230 indigenous groups known from Brazil in 1900 had died out by 1957 (Ribeiro 1957). Summarizing the situation, the anthropologist John H. Bodley estimates that 30–50 million tribal people have died over the past 150 years as a direct result of the spread of industrial civilization (Bodley 1982, 40).

Religious adherence is another indicator of cultural diversity for which at least rough global statistics can be estimated. Currently, there are some 10,000 active religions, 6,000 of which are termed "ethnoreligions," that is, religions restricted to a single tribal group (Barrett, Kurian, and Johnson 2001, 2:4). Over the course of history, the total number of religions is very large, perhaps on the order of 100,000 (Wallace 1966, 3).[45] Within faiths, especially polytheistic belief systems, the number of spiritual beings is astounding. Some 3,300 gods existed in ancient Mesopotamia alone, and (according to David Hume) in the Athens of Hesiod's time there were said to be 30,000 deities. A list of the world's gods over time "would probably form a list as bulky as the telephone directory of a large city" (Hick 1989, 234). Most people now follow one of the "world religions": Christianity, Islam, Hinduism, Buddhism, or Judaism. In 1900, 68 percent of the world's population was affiliated with one of them; by 2000, 72 percent were (Table 2.5). The increase can be attributed to the strong growth of Islam, which alone among the five saw its proportion rise significantly. In numerical terms, almost every major religious category grew during the 20th century, but much of this is a function of the massive global population growth during the period. In percentage terms, two categories of religions have fared poorly. Chinese folk religions, which claimed 23.6 percent of the world's people in 1900, declined steeply after the advent of Communism, falling to just a 6.4 percent share in 2000. This percentage decline may well represent a commensurate loss of global religious diversity. So too, probably, does the drop in percentage adherence to tribal religions (which include forms of shamanism and animism). Part of the decline can be laid at the doorstep of governments that have persecuted shamanistic faiths (Bodley 1982, 115–117). In 1910, the World Missionary Conference predicted that native religions would be gone within a generation (Barrett 1982, 5). Yet, against all odds, tribal religions have persisted. Their survival in the face of intense proselytizing and outright suppression is a remarkable testament to the durability of cultural beliefs.

All this notwithstanding, the most important trend in religious adherence during the 20th century was the tremendous jump in both the numbers and the percentage of the world's population that became secularized, whether as agnostics or atheists. As of 2000, about 15 percent claimed no religion, up from a negligible proportion in 1900. According to the *World Christian Encyclopedia*, most of the newly secularized had Christian parents, grandparents, or great-grandparents. Such spectacular growth was totally unexpected, for "no Christian strategist in 1900 had envisaged such a massive rate of defection from Christianity within its 19th-century heartlands" (Barrett, Kurian, and Johnson 2001, 1:5).

Beyond the rise of secularization and the percentage declines in Chinese folk and tribal religions, it is hard to draw any meaningful conclusions about trends

Table 2.4

Enumeration of World Population by Ethnolinguistic Families, 1900–1980

Race and Ethnolinguistic Family	Population, 1900 (Thousands)	Population, 1980 (Thousands)	Change Relative to World Increment, 1900–1980 (Index = 1.0)
Australoid			
Ainu/Aborigine	58	348	2.9
Australian Aborigine	41	167	1.8
Fijian	54	284	2.5
Melanesian	585	1,825	1.2
Mon-Khmer	3,541	13,758	1.7
Munda-Santal	3,508	10,593	1.2
Negrito	78	522	3.3
Papuan	1,329	4,109	1.2
Pre-Dravidian	2,995	9,068	1.2
Vedda	5	<1	—
Capoid			
Khoisan	50	251	2.4
Pygmy	84	295	1.5
Caucasian			
Albanian	1,428	4,230	1.1
Arab	33,929	142,374	1.8
Armenian	2,495	5,813	0.7
Assyrian	123	637	2.4
Baltic	2,186	4,826	0.7
Basque	555	1,017	0.4
Berber	3,151	12,904	2.4
Blanco (White)	12,101	64,883	2.5
Branco (White)	9,553	67,089	3.5
Caucasian	3,352	7,285	0.6
Celtic	6,619	9,480	0.2
Cushitic	5,824	23,585	1.7
Dravidian	58,557	179,037	1.2
Ethiopic	2,856	11,254	1.7
Germanic	163,617	368,673	0.7
Greek	4,017	11,449	1.1
Iranian	17,794	67,596	1.6
Jewish	4,852	15,171	1.2
Latin	112,694	199,749	0.4
Mestiço	2,158	15,167	3.5
Mestizo	23,776	113,396	2.2
North Indian	225,669	687,054	1.2
Maltese	216	390	0.4
Slav	142,235	282,021	0.5
Mongolian			
Altaic	42,669	108,803	0.9
Central Amerindian	5,225	25,211	2.2
Chinese	454,047	886,376	0.9

Table 2.4 *continued*

Race and Ethnolinguistic Family	Population, 1900 (Thousands)	Population, 1980 (Thousands)	Change Relative to World Increment, 1900–1980 (Index = 1.0)
Eskimo-Aleut	15	74	2.3
Eurasian	53	222	1.8
Euronesian	257	1,728	3.3
Indo-Malay	48,806	217,847	2.0
Japanese	44,877	118,473	0.9
Korean	13,377	58,179	1.9
Miao-Yao	2,567	5,242	0.6
Micronesian	54	194	1.4
Northern Amerindian	483	1,683	1.4
Paleoasiatic	20	26	0.1
Polynesian	259	1,037	1.7
Southern Amerindian	4,233	19,896	2.1
Tai	13,209	58,181	2.0
Tibeto-Burmese	17,435	50,510	1.1
Uralian	13,338	23,648	0.4
Viet-Muong	10,227	43,613	1.9
Negro			
Bantoid	9,391	43,657	2.1
Bantu	31,076	134,643	1.9
Eurafrican	869	4,049	2.1
Guinean	10,179	48,139	2.1
Hausa-Chadic	4,799	20,789	1.9
Kanuri	1,094	4,621	1.8
Nilotic	3,802	17,074	2.0
Nuclear Mande	1,684	8,404	2.3
Peripheral Mande	750	3,325	2.0
Songhai	311	1,677	2.5
Sudanic	4,498	14,207	1.2
Afro-American			
Dutch-speaking	97	604	3.0
English-speaking	11,378	33,465	1.1
French-speaking	2,065	6,007	1.1
Portuguese-speaking	5,941	41,743	3.5
Spanish-speaking	2,760	20,163	3.7
World total	1,619,887	4,373,918	1.0
5 largest families in 1900 (% of world total)	1,098,262 (68%)	2,423,873 (55%)	0.7

Sources: 1900 and 1980 populations figures from Barrett 1982, 786–789. Index calculated by the author, as follows. (1) The total world population increment from 1900 to 1980 was 2,754,031 thousands. (2) This increment represents a 170% increase over the 1900 world population: 2,754,031 / 1,619,887 = 1.70. Thus 170% becomes the index figure (represented by 1.0), i.e., the world growth average for the period 1900–1980. (3) The calculation is repeated for each ethnolinguistic family to obtain its percentage increase for the period 1900–1980 (except for Vedda, the only ethnolinguistic group whose population declined). (4) This percentage is divided by 1.70 to get a score relative to the world index figure. A score of <1.0 indicates that the ethnolinguistic family grew more slowly than the world average; a score of <1.0, that it grew faster.

Table 2.5

Enumeration of World Population by Religion, 1900–2000

Religion	1900 Adherents (Millions)	%	1970 Adherents (Millions)	%	1990 Adherents (Millions)	%	2000 Adherents (Millions)	%	Number of countries	%
Baha'ism	<0.1	<0.1	2.6	0.1	5.6	0.1	7.1	0.1	218	91.5
Buddhism	127.0	7.8	233.4	6.3	323.1	6.1	359.9	5.9	126	52.9
Chinese folk religions	380.0	23.5	231.8	6.3	347.6	6.6	384.8	6.4	89	37.3
Christianity	558.1	34.5	1,236.3	33.5	1,747.4	33.2	1,999.5	33.0	238	100.0
Confucianism	0.6	<0.1	4.7	0.1	5.8	0.1	6.2	0.1	15	6.3
Ethnoreligions[a] (Tribal religions)	117.5	7.3	160.2	4.3	200.0	3.8	228.3	3.8	142	60.1
Hinduism	203.0	12.5	462.5	12.5	685.9	13.0	811.3	13.4	114	47.8
Islam	199.9	12.3	553.5	15.0	962.3	18.3	1,188.2	19.6	204	85.7
Jainism	1.3	0.1	2.6	0.1	3.8	0.1	4.2	0.1	10	4.2
Judaism	12.2	0.8	14.7	0.4	13.1	0.3	14.4	0.2	134	56.3
Mandeanism	<0.1	<0.1	<0.1	<0.1	<0.1	<0.1	<0.1	<0.1	2	0.8
New religions	5.9	0.4	77.7	2.1	92.3	1.8	102.3	1.7	60	25.2
Shintoism	6.7	0.4	4.1	0.1	3.0	0.1	2.7	<0.1	8	3.3
Sikhism	2.9	0.2	10.6	0.3	19.3	0.4	23.2	0.4	34	14.2
Spiritists	0.2	<0.1	4.6	0.1	10.1	0.2	12.3	0.2	55	23.1
Taoism	0.3	<0.1	1.7	<0.1	2.4	<0.1	2.6	<0.1	5	2.1
Zoroastrianism	0.1	<0.1	0.1	<0.1	1.9	<0.1	2.5	<0.1	24	10.0
Other religions	<0.1	<0.1	0.7	<0.1	0.9	<0.1	1.0	<0.1	76	31.9
Nonreligious[b]	3.0	0.2	532.0	14.4	707.1	13.4	768.1	12.7	236	99.1
Atheism	0.2	<0.1	165.4	4.5	145.7	2.8	150.0	2.5	161	67.6
World total	1,619.6	100	3,696.1	100	5,266.4	100	6,055.0	100	238	—

Note: Totals may not add due to rounding.

[a]Includes animism, shamanism, polytheism, and other so-called primal religions.

[b]Includes agnosticism, secular humanism, etc.

Source: Barrett, Kurian, and Johnson 2001, 1:4.

in religious diversity from the data in Table 2.5. For one thing, the numbers are extremely fluid. Each year millions of people change religious affiliations or drop them altogether. In times of political upheaval, the change is even greater; for instance, after the collapse of the Soviet Union, the Russian Orthodox Church rebounded strongly after decades of Communist oppression (Barrett, Kurian, and Johnson 2001, 1:626). Second, it is possible that any losses of diversity may have been offset by the worldwide rise in new religions, whose adherents increased from 6 million in 1900 to 102 million in 2000. Scholars of religion recognize two prevalent processes that create new belief systems: *syncretism*, the blending of elements of two traditions, and *bricolage*, the amalgamation of many bits and pieces of diverse cultural systems (Taylor 1997, 198, citing Stewart and Shaw 1994, 20). The mass migrations, vastly improved communications, and increasing (though by no means universal) civil and political freedoms of the 20th century have all encouraged the development of new religions. In countries with a laissez-faire policy toward faith, the religious experience has in all likelihood diversified over the past hundred years. For example, a recent encyclopedia lists more than 1,600 churches now active in the United States and Canada, many of them formed in this century (Melton 1989).

LANGUAGE: THE BEST PROXY

What we have discussed so far are only a few of the most prominent indicators of global cultural diversity. One could, theoretically, expand the list ad infinitum. In reality, however, there are very few indicators that are universally applicable and for which reasonable global estimates exist (or could conceivably exist). Of the available choices, one stands out from the rest. Language use is by far the best gauge of the overall trend in cultural diversity. Or, to be precise, the best gauge is linguistic diversity: the range of variation (syntactical, grammatical, phonical, lexical, etc.) exhibited across languages, dialects, pidgins, creoles, registers, and other forms of speech.[46]

Like biodiversity, linguistic diversity can be measured on three fundamental, hierarchically related levels. Structural linguistic diversity, the lowest level, has to do with the amount of disparity exhibited by the structural features and types within a language or population of languages (Nichols 1992, 237). Lineage diversity, the highest level, refers to the "number of discrete lineages and the extent to which individual lineages have branched out. The more branches or lineages, the greater the diversity" (Nichols 1992, 232). Diversity in language, the middle level, most properly refers to the range of variation between individual languages across lineages. This is difficult to measure because, as any linguist can tell you, there are raging disagreements over how to classify languages. So, just as species richness serves as the de facto standard of biodiversity, language richness—the number of distinct languages in use—stands in for all of linguistic diversity, becoming, by extension, the most effective measure of cultural diversity.

Admittedly, language is not equivalent to culture, as is often implied in loose or careless usage. There are numerous instances where the same language is spoken by groups with otherwise radically different cultural practices. But it *is* the carrier of many cultural differences, and many people consider it to be emblematic of distinctive world views.[47] Of all the indicators of cultural diversity, it offers the best chance of making a comprehensive and comprehensible division of the world's peoples into constituent groups based on a single aspect of culture. Unlike religion, language extends into every corner of existence, the mundane as well as the sacred, and is used by virtually everyone every day. Unlike ethnicity, language affiliations are comparatively straightforward (though far from clear-cut). Ethnic identification involves a complex reciprocal arrangement in which an individual associates her- or himself with a group, and the group in turn accepts the association. Such an arrangement may occur along with language affiliation, but it need not. One can be recognized as a fluent speaker of Japanese on the basis of having mastered the language without making any pretense to "being Japanese" in a cultural or ethnic sense.

Language affiliations are more comprehensive and readily quantifiable than either religious or ethnic ones. We must be cautious not to overstate matters here, however. It certainly is possible to aggregate language-use statistics on a global basis, and, as we shall shortly see, it has been attempted (against long odds) by the compilers of reference works. Nevertheless, we remain a long, long way from having really accurate figures on global language use.[48] As an ideal, if we could on a single day ask each person on the planet to name his or her native language (that is, his or her mother tongue of origin, defined below), the results would be an approximation of the world's cultural diversity at that moment.

I emphasize *approximation*. Even if it were possible, our imaginary census would not capture the nuances of language use, let alone those of culture in general. For example, if "mother tongue" is defined as a person's "first language," that through which he or she communicates most effectively (Grimes

Table 2.6
Skutnabb-Kangas's Definitions of "Mother Tongue"

Criterion	Definition
1. Origin	The language one learned first (i.e., the language one has established the first long-lasting verbal contacts in)
2. Identification	
Internal	The language one identifies with / as a native speaker of
External	The language one is identified with / as a native speaker of, by others
3. Competence	The language one knows best
4. Function	The language one uses most

Source: Skutnabb-Kangas 2000, 106.

1992b, vi), then it follows that even a person raised in a bilingual or multilingual household can have only one mother tongue. However, the two criteria embedded in this definition—that of priority, as in the language one uses first if given a choice, and that of competence, as in the language one communicates most effectively in—are not the same and do not necessarily coincide. The work of the linguist Tove Skutnabb-Kangas supports this conclusion. Table 2.6 gives her four criteria for defining "mother tongue." Skutnabb-Kangas herself has two mother tongues by origin, having learned Swedish and Finnish more or less equally from earliest childhood; two by identification (Finnish and Finland's version of Swedish, which differs from the "standard" Swedish spoken in Sweden proper); two by competence (Swedish and Finnish again); and two by function (English and Danish, in which she now most often works as an academic and a resident of Denmark, respectively). By these criteria, one need not have a single mother tongue—a conclusion which undercuts the assumption behind our one-off census. And in fact Skutnabb-Kangas's more finely shaded conception of "mother tongue" better reflects the experience of millions of people whose daily lives include the equally effective use of two (or more) languages, usually in different spheres: one at home, another at work or the market, perhaps a third at worship, and so on. Such multilingualism may well have been the norm throughout history (Grimes 1995).

All this notwithstanding, we could (as a querying technique in our imaginary census) get most, if not all, multilingual people to identify themselves with a single mother tongue by (1) allowing them to select one of Skutnabb-Kangas's four criteria as being most appropriate to their situation, and then (2) prioritizing the language choices within that criterion.[49] When done, we would have established a one-to-one equivalence between the cumulative number of mother-tongue speakers of the world's languages and the global population. That, in turn, would result in the comprehensive approximation of cultural diversity we are seeking.[50]

THE HISTORY OF LANGUAGE EXTINCTIONS: SOME SPECULATIONS

As was done above with species, let us now briefly consider the deep history of language genesis and extinction. In contrast to the rich (if incomplete) physical record attesting to the past of biological diversity, there are no ancient "fossils" of language. Attempts to go back tens of thousands of years and reconstruct proto-languages (e.g., Ruhlen 1994) have met with skepticism and even derision from many professional linguists. Conventional linguistic wisdom holds that 8,000–10,000 years is the furthest back anyone can reliably trace language change using a comparative method (Pinker 1994, 256–259). The timeline of reliable knowledge about language being much shorter than that of species, linguists are left to hypothesize about the overall course of language growth and decline throughout history, and more specifically about whether languages have undergone any mass extinctions in the distant past comparable in proportion to those found in nature.

However, by using a series of inferences it is possible to roughly estimate the background rate of language extinctions through historical time. The linguist Michael Krauss reasons that the highest number of languages must have been reached just before the onset of the Neolithic period:

> At that point, about 10,000 years ago, human population had spread over the globe, numbering most likely between 5 and 10 million . . . .[51] It's hard to imagine that there existed then any large societies or single languages with sizable numbers (10,000 or more) of people/speakers; only after the invention of food production were people able to live relatively densely in large societies. Before these larger-scale societies began to expand at the cost of the hunter-gatherers, average language size must have been much closer to the median (500 to 1,000 per language) than today. So with global population at 5 to 10 million, the extremes for number of languages would be from 5,000,000 ÷ 1,000 or 5,000, to 10,000,000 ÷ 500, or 20,000, with a mid-range of 10,000 to 15,000. (Krauss 1995a, 5)

Although such calculations must always be very imprecise, Krauss's reasoning is solid. Today the world's language demographics are characterized by extremes of disparity. There are thousands of languages with very few mother-tongue speakers and only about 300 with 1 million or more (Harmon 1995). At present, the *median* language size is 5,000 speakers, with the *average* being 900,000—a ratio of 1:180.[52] Such a ratio could not have prevailed in a world of 5–10 million people, unless we are prepared to accept a scenario in which many languages were spoken by literally only a handful of people. Therefore, Krauss is correct: 10,000 years ago the average language size had to be much closer to the median than it is today. The rest of his calculations flow from that, so I believe his mid-range estimate of 10,000–15,000 languages is reasonable.

Of course, plugging in even slightly different estimates will change the result considerably. The anthropologist Jane H. Hill accepts John Robb's proposition (Robb 1993) that "the apex of human linguistic diversity may have been reached in the Neolithic, with a sociolinguistic milieu perhaps similar to that among contemporary horticulturalists in the Upper Amazon or Highland New Guinea being found planetwide. In these regions hundreds of small languages are found in relatively small areas, with speaker numbers ranging from a few hundred to perhaps 10,000" (Hill 2001, 175). If we assign a value of 500 to Hill's locution "a few hundred," her language size range is 500–10,000. Following Krauss's calculations thenceforward, we end up with 500–20,000 for the total number of languages. If we change the equation further and accept the demographer Joel E. Cohen's estimate of 2–20 million people on Earth in 10,000 B.C. (Cohen 1995, 77, 400, 453) and apply Krauss's calculations, we get a range of 2,000–40,000 languages. Tore Janson (cited in Skutnabb-Kangas 2000, 216) estimates that there was one language for every 1,000–2,000 people at that time; combining that with Cohen's population range, we get 1,000–20,000 languages. Using similar calculations, Daniel Nettle gives a range of 1,667–9,000 languages just before the onset of the Neolithic (Nettle 1999a, 102). We could go on (and apply more sophisticated guesswork to the trajectory of language formation and extinction, as in Robb 1993), but the general picture is clear: the middle ground of these estimates is 10,000–25,000 languages.

If we accept a count of about 6,800 currently living languages (Grimes 2000a, 846)—which, it should be cautioned, is by no means a settled estimate[53]—this leaves us with a net loss of 3,200–18,200 languages over 10,000 years[54] or an average background extinction rate of 0.32–1.82 languages per year.[55] For all practical purposes, and to avoid giving an appearance of precision that is completely unwarranted, we can call it 1 language per year. This rough estimate of the background rate of language extinctions is in line with that of the biological extinctions, which, as you will recall, ranged from <1 to 4 species per year.

PROSPECTS FOR LANGUAGES IN THE COMING CENTURY

More and more, linguists are realizing that the 21st century will very likely witness the extinction of a great number of languages around the world. This consensus has arisen largely out of the shared experience of those working on individual endangered languages, or groups of them, from various regions. As with biodiversity, the evidence for an impending mass extinction derives not from accurate knowledge of the global situation, which does not exist, but from the relatively few smaller languages for which linguists have obtained reasonably detailed demographic information.

The crucial question is whether a language is moribund. This occurs when it is no longer being passed on to the younger generations of the speech community. The failure of intergenerational transmission in a language is analogous to the loss of reproductive capacity in a species (Krauss 1992, 4). In a quantitative study of language vitality and moribundity (Norris 1998), Statistics Canada used 1996 census responses to calculate an "index of continuity" and an "index of ability" for the country's indigenous languages. The index of continuity measures a language's vitality by comparing the number who speak it at home with the number who learned it as their mother tongue of origin. In this index, a 1:1 ratio is scored at 100 and represents a perfect maintenance situation in which every mother-tongue speaker keeps the language as a home language. Any score lower than 100 indicates a decline in the strength of the language. The index of ability compares the number who report being able to speak the language (at a conversational level) with the number of mother-tongue speakers. Here, a score of over 100 indicates that an increment of people have learned it as a second language, and may suggest some degree of language revival (Norris 1998, 10).

Table 2.7 shows the main results of the study. All the elements of a thorough moribundity index are here: the size of the speaking population, indices of continuity and rejuvenation, and the average age of the speakers. By combining the two indices and adjusting the result by judiciously weighting the other factors, one could derive a quantitative measure of a given language's vitality or lack thereof. Doing this on a global scale would require every country to conduct a census as thorough as Canada's (which nonetheless still suffers from incomplete enumeration of some native reservations).

Most moribundity information comes from anecdotal reports that young people no longer speak the language in certain social situations, or that they have given it up altogether. We know, for example, that the great majority of indigenous languages in North America are in trouble. Taking Canada and the United States together, probably 80 percent are moribund (Krauss 1992, 5) and, if current trends continue, the general outlook for the continued survival of these languages is very poor (Kinkade 1991; Zepeda and Hill 1991). The Statistics Canada report just cited found that only 3 of the country's 50 native languages have speaking populations large enough to keep them secure over the long run, and the index of continuity for all native languages declined almost 15 percent between 1981 and 1996 (Norris 1998, 8, 12). Leanne Hinton's review of the native languages of California paints a similar picture: she surveyed 50 contemporary languages in the state (at least 50 others existed before Europeans arrived) and found that more than 15 are recently extinct, many others have fewer than 10 speakers (all elders), and only two or three have as many as 150 or 200 speakers (Hinton 1994, 27–33). Along the temperate rainforest coast, stretching from northern California to Alaska, of the 68 native language groups present at European contact, 26 are extinct, 18 are spoken by fewer than 10 people, and only 8 are spoken by more than 100 (Kellogg 1995, 6–7). In Mexico, "the danger of extinction exists in all areas" where indigenous languages are found (Garza Cuarón and Lastra 1991, 98).

I could go on with examples from Aboriginal Australia, from Siberia, from southern Africa, even from Europe—where no regional language, from the Sámi tongues to Basque, from Frisian to Welsh, is completely secure. But what can we say about the global situation? Both the empirical and anecdotal evidence point to an impending extinction crisis. Unfortunately, we do not have a systematic synthesis of global endangerment data that would give an accurate estimate of how great the magnitude of the crisis will be. As we have seen, under the best of conditions it would take a massive effort to gather the needed demographic data. Because many countries actively seek to destroy the minority languages within their boundaries, the political climate for surveying language use is more hostile than favorable, and will likely remain so for years to come.

The closest thing we now have to a global synthesis of the status of languages is *Ethnologue: Languages of the World*, published every four years by the

Table 2.7
Status of Native Languages in Canada, 1996

Language	Mother-Tongue Population	Index of Continuity	Index of Ability	Average Age of Mother-Tongue Speakers
Algonquian family	146,635	70	117	30.9
Cree	87,555	72	117	30.2
Ojibway	25,885	55	122	36.2
Montagnais-Naskapi	9,070	94	104	25.2
Micmac	7,310	72	111	29.9
Oji-Cree	5,400	80	114	26.3
Blackfoot	4,145	61	135	39.7
Attikamek	3,995	97	103	21.9
Algonquin	2,275	58	119	30.7
Malectite	655	37	148	44.0
Other Algonquian	350	40	159	52.2
Inuktitut family	27,780	86	109	23.9
Athapaskan family	20,090	68	117	32.5
Dene	9,000	86	107	24.8
South Slave	2,620	55	124	37.8
Dogrib	2,085	72	118	29.8
Carrier	2,190	51	130	41.4
Chipewyan	1,455	44	128	40.2
Athapaskan	1,310	37	129	44.7
Chilcotin	705	65	130	37.0
Kutchin-Gwich'in (Loucheux)	430	24	114	53.1
North Slave (Hare)	290	60	116	39.1
(Dakota) Siouan family	4,295	67	111	31.9
Salish family	3,200	25	132	48.7
Salish	1,850	24	130	49.7
Shuswap	745	25	134	46.3
Thompson	595	31	135	48.6
Tsimshian family	2,460	31	132	48.0
Gitksan	1,200	39	123	45.2
Nishga	795	23	146	47.5
Tsimshian	465	24	132	55.9
Wakashan family	1,650	27	118	51.3
Wakashan	1,070	24	129	53.0
Nootka	590	31	99	48.1

Table 2.7 *continued*

Language	Mother-Tongue Population	Index of Continuity	Index of Ability	Average Age of Mother-Tongue Speakers
Iroquoian family	590	13	160	46.5
Mohawk	350	10	184	46.1
Iroquoian	235	13	128	47.0
Haida family	240	6	144	50.4
Tlingit family	145	21	128	49.3
Kutenai Family	120	17	200	52.3
Other native languages	1,405	28	176	47.0
Total	208,610	70	117	31.0

Source: Modified from Norris 1998, 13. Data for the Iroquoian family are not particularly representative due to incomplete enumeration of reserves. Other languages may also be affected by incomplete enumeration.

Summer Institute of Linguistics (SIL), and now available on the Internet.[56] *Ethnologue* has been called "by far the best single source available" for the numbers of languages and their speakers globally (Krauss 1992, 4n 1) and is the most widely cited authority. It draws on a multitude of sources (such as field reports by linguists as well as the published literature) to produce a country-by-country listing of the world's languages. The most recent (14th) edition of *Ethnologue* lists 6,809 living languages (Grimes 2000a, 846).[57]

Although *Ethnologue* is a monumental achievement, an indispensable reference, its database has many shortcomings. As one must expect, its sources are wildly uneven, ranging from citations out of popular almanacs down to field reports from trained linguists. Some of the data are very old. Furthermore, it is possible that *Ethnologue* overstates the global number of languages by listing some dialects as separate languages—or understates the true number by omitting many deaf languages (Adelaar 1991, 53; Maffi 1999, 22; Skutnabb-Kangas 2000, 30).[58] Because SIL is a Christian missionary organization (one whose role in language politics and acculturation is itself very controversial), the database places a great deal of emphasis on the status of Bible translations in various languages. This, along with the country-by-country data reporting format, makes the database difficult to work with in terms of extracting information on language endangerment, though the editors hope to include better data in future editions (Grimes 2000a, vii).

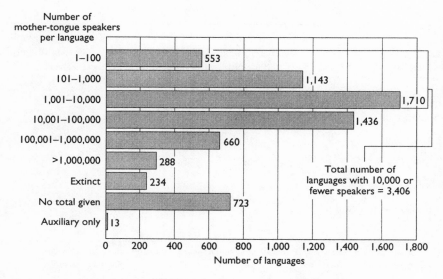

Figure 2.4. Size classification of world's languages by number of mother-tongue speakers: numerical totals (*n* = 6,760). Source: Harmon 1995, 8, using data derived from Grimes 1992b. Reprinted with permission of the *Southwest Journal of Linguistics*.

To get a sense of the status of the world's languages, in a recent study (Harmon 1995) I analyzed the 1992 *Ethnologue* data (Grimes 1992b) by creating an endangerment-oriented "shadow" database of all 6,760 languages (6,526 living and 234 extinct) listed in that edition.[59] It is worth stressing that I handled the data as conservatively as possible, so the following figures are the most optimistic one can derive from *Ethnologue*. In all likelihood, they understate the severity of the global language endangerment situation. Here are the three main insights from the study.

1. Most languages are small.   Figure 2.4 divides the world's languages by size: 1–100 mother-tongue speakers, 101–1,000, and so on. The key point of this graphic is that most languages fall into the smaller categories. This is reinforced by Figure 2.5, which considers living languages only. Of these, about 52 percent are spoken by 10,000 or fewer people. This is a minimum figure; undoubtedly the real percentage of languages in the "10,000-and-under" category is even higher. To see why, return for a moment to Figure 2.4; there, we find 723 languages for which *Ethnologue* gives no indication of a total. Of these, 666 are endemic to a single country. This automatically predisposes the group

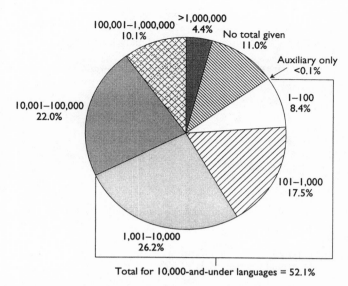

Figure 2.5. Proportion of world's living languages by size category (*n* = 6,526). Source: Harmon 1995, 9, using data derived from Grimes 1992b. Reprinted with permission of the *Southwest Journal of Linguistics*.

toward the low end of the size scale. I think it is not unrealistic to hazard a guess that 75 percent of these 723 "no total given" languages have no more than 10,000 mother-tongue speakers. If that is so, then the actual global proportion of "10,000-and-under" languages would approach 60 percent. Thus, between 52 percent and 60 percent of the world's living languages—something like 3,400–4,000 languages—are spoken by no more than 10,000 people. Many of these are, of course, indigenous languages.

   2. Most languages are endemics.   Given that so many are spoken by small numbers of people, it is no surprise that a little over 80 percent of the world's languages are reported as being endemic to a single country. So, on top of being vulnerable because of their small size, most of the world's languages are susceptible to the vagaries of a single government's policy. This suggests that government support of or antipathy toward minority languages will play a large role in determining language extinctions in the years to come. A more insidious threat comes from unique social pressures within individual countries that favor certain languages. These factors appear to be widespread around the world.

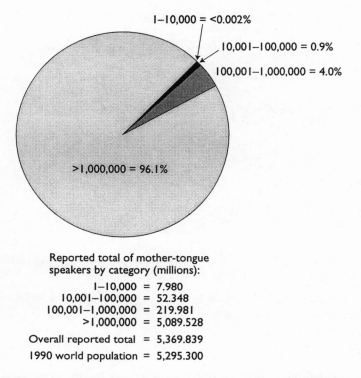

Reported total of mother-tongue
speakers by category (millions):

1–10,000	=	7.980
10,001–100,000	=	52.348
100,001–1,000,000	=	219.981
>1,000,000	=	5,089.528
Overall reported total	=	5,369.839
1990 world population	=	5,295.300

Figure 2.6. Proportion of 1990 world population by size category (living languages with reported totals only; $n = 5,509$). Source: Harmon 1995, 10, using data derived from Grimes 1992b. Reprinted with permission of the *Southwest Journal of Linguistics*.

3. The largest languages predominate.   Numerous as they are, small languages are the mother tongues of only about 8 million people, or far less than 1 percent of the world's population (Figure 2.6). The 300 largest languages—basically those having over 1 million speakers each—utterly dominate the world linguistic scene. The figures in *Ethnologue* suggest they are spoken by something like 5 billion people.[60] Let us say, to simplify things, that nine out of ten people speak one of 300 languages as a mother tongue. Thus we can make a startling comparison: namely, that 52–60 percent of all languages are spoken by far less than 1 percent of the people, while less than 5 percent of all languages are spoken by 90 percent of the people. Indeed, almost half the people in the world speak one of the ten largest languages as their mother tongue. Figure 2.7 brings the entire picture into focus. The overwhelming domi-

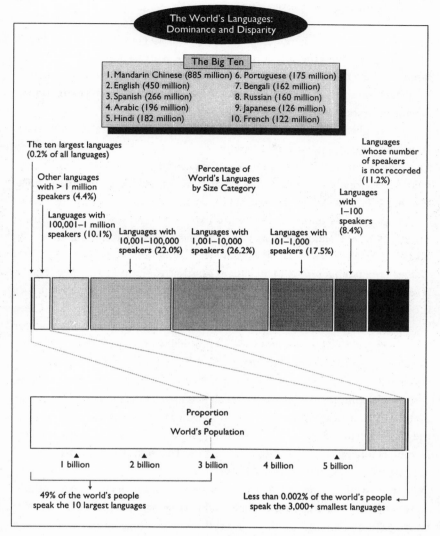

The World's Languages:
Dominance and Disparity

The Big Ten

1. Mandarin Chinese (885 million)
2. English (450 million)
3. Spanish (266 million)
4. Arabic (196 million)
5. Hindi (182 million)
6. Portuguese (175 million)
7. Bengali (162 million)
8. Russian (160 million)
9. Japanese (126 million)
10. French (122 million)

The ten largest languages
(0.2% of all languages)

Other languages
with > 1 million
speakers (4.4%)

Languages with
100,001–1 million
speakers (10.1%)

Percentage of
World's Languages
by Size Category

Languages with
10,001–100,000
speakers (22.0%)

Languages with
1,001–10,000
speakers (26.2%)

Languages with
101–1,000
speakers (17.5%)

Languages
whose number
of speakers
is not recorded
(11.2%)

Languages
with
1–100
speakers
(8.4%)

Proportion
of
World's Population

▲ 1 billion   ▲ 2 billion   ▲ 3 billion   ▲ 4 billion   ▲ 5 billion

49% of the world's people
speak the 10 largest languages

Less than 0.002% of the world's people
speak the 3,000+ smallest languages

Figure 2.7. The world's languages: dominance and disparity. The upper bar represents the proportion of the world's languages by size category; the lower bar, the global population divided into mother-tongue speakers. The dominance of the ten largest languages (top box) is shown by the fact that, while they account for just 0.2% of all languages spoken (narrow slice at far left of upper bar), they comprise 49% of the world's people (lower bar). Contrast this with the number of languages spoken by 10,000 or fewer people: these smallest languages account for well over half the world's total (rightmost four slices of the upper bar), but far less than 1% of the world's population (narrow slice at far right of lower bar). Source: Harmon 1998b, 6. Reproduced from D. Harmon, "Sameness and silence: Language extinctions and the dawning of a biocultural approach to diversity," *Global Biodiversity*, 8(3), published by the Canadian Museum of Nature, Ottawa, Ontario, Canada.

nance of the "big ten" languages, the disparity between the great number of small languages and the tiny proportion of the world's population they represent, and the multifarious sociopolitical pressures bearing down on minority languages are three macro-factors that point to a probable mass extinction of languages in the coming century.

What might the ultimate toll be? Given the quality of data available to us at present, any figures we quote or come up with must be classed as informed speculation. Michael Krauss thinks that perhaps 50 percent of the world's languages are already moribund, that 20–50 percent will likely become extinct, and that as many as 90 percent might become extinct (or have such a tiny number of speakers as to be committed to extinction) during the next century (Krauss 1992, 4; Krauss 1995b, 4). Other linguists agree. Peter Mühlhäusler has stated (1995, 156), without elaboration, that more than 95 percent of all languages are endangered. R.M.W. Dixon says flatly: "Nothing can be done to reverse or arrest the continuing reduction in the number of distinct languages spoken in the world, although the rate of reduction could be slowed. There were originally perhaps four to five thousand separate languages; by the year 2100 there will be many fewer—perhaps only a few hundred" (1991, 234). This statement implies an overall reduction of 75 percent or more. Dixon's gloomy assessment of total irreversibility is softened only slightly by Daniel Nettle, who, while pointing to increasing interest in endangered languages as evidence that the current wave of kill-offs is not necessarily unstoppable, still considers the current crisis "particularly grave." "Most of our human heritage is disappearing before our eyes," he avers, and suggests that overall reductions of 83–95 percent are real possibilities (Nettle 1999a, 112–114).

My review of *Ethnologue* data supports the magnitude of these speculations. As we have seen, 52–60 percent of the world's languages are spoken by no more than 10,000 people. All of these, even those now considered vigorous and viable, are at significant risk simply because of their small size. Economic dislocations, shifts in government policy, a new access road, crop failures, natural disasters, a sudden influx of immigrants—any number of unpredictable events could uproot a stable culture and tip a small language toward extinction. The larger economic and sociopolitical trends—the decrease in physical isolation because of better transportation, the growing reach of telecommunications, the heightened place of the "information industry" in an increasingly fluid and interlinked world economy, the dominance of English as the language of advancement in business and academia, and the active repression of minority languages by many governments, to name just a few—are not likely to subside any time soon. Because the pressures on small languages will only increase, my guess is that a very high percentage—let us say 80 percent—of these "10,000-and-under" languages will cease to be spoken as mother tongues, or at least will become irreversibly moribund, in the century to come. If so, that would mean the extinction of roughly 2,800–3,200 languages, or 43–49 percent of all living languages. Yet undoubtedly some larger languages will also die out. Taking the 2,800–3,200 figure as a base, I think it is reasonable to predict that at least 50 percent of the world's languages will become extinct or irreversibly moribund as mother tongues during the next century.

This figure I consider to be a probable *minimum*. My speaker totals err on the high side; the actual number of speakers of many languages may be fewer than is recorded in my database. It is quite conceivable that more than 80 percent of the "10,000-and-under" languages will become extinct as mother tongues, and the number of extinctions in the larger size categories may be greater than any of us now imagine. Recall that nine out of ten of the world's people speak one of about 300 large languages. If, for the sake of argument, we assume that these largest languages are not in danger of extinction,[61] then the question becomes: How many of the remaining 10 percent of the people will switch to a larger language? The pressure to do so will be intense. Another way to approach the issue is to reason backwards from a plausible result. Even if as many as 90 percent of the 6,809 living languages reported in the 2000 edition of *Ethnologue* were to become extinct or committed to extinction (by becoming irreversibly moribund) during this century, we would still be left with 680 languages—more than twice the number of large languages now existing. As a worst-case scenario of language loss, this seems entirely plausible.

A caution, though: language extinction is an extremely complicated phenomenon. Raw numbers of speakers are a starting point, not the end. For one thing, as Joseph Grimes has pointed out, the size at which a language becomes critically endangered seems to vary from region to region: for example, a language with 500 speakers may be in trouble in Africa but not in the Pacific because of differing social and political conditions (Grimes 1995, 8). Likewise, many other circumstances related to extinction cannot be inferred from numbers alone: the language community's own attitudes toward its language, either positive or negative, and the plethora of related sociolinguistic factors; government attitudes and policies toward minority languages; and the presence or absence of intervention programs to preserve or perpetuate languages (for examples, see Zepeda and Hill 1991, 142–147; Hinton 1994, 235–247; Maffi and Skutnabb-Kangas 1999, 37–45).

And, without giving the subject the attention it deserves, it should be mentioned that the death and extinction metaphor—while undeniably useful with respect to language—must be handled carefully (cf. Hoenigswald 1989, 347–348). For example, languages can be functionally dead long before the last native speaker dies. They can also live on post-extinction in a transmuted form: no longer as mother tongues, but as languages of heritage.[62]

THE EXAMPLE OF MANX

The example of Manx shows this clearly. Once spoken on the Isle of Man (located in the Irish Sea midway between Britain and Ireland), Manx became extinct as a mother tongue in 1974 when the last native speaker, a man named Ned Maddrell, died. But its death was in the making more than a hundred years before. A young clergyman, Henry Jenner, visited Man in 1874 and 1875 expressly to ascertain the status of Manx. "During the whole of my tour I only met with one person who could not speak English, though I went into a good many cottages on various pretexts of resting, asking the way, etc., so as to find some such person if possible." Even then, Jenner felt that Manx had already "certainly received its death-blow" when it "ceased to be taught in schools and its use was discontinued in most of the churches. Those who speak it now are all of them old people, and when the present generation has grown up and the older folk have died off, it will cease to be the mother tongue of any Manxmen. . . . How long these will last it is hard to say; but there is a decided feeling on the part of the people, especially among the Manx speakers themselves, that the language is only an obstruction, and that the sooner it is removed the better." In a series of visits from 1886 to 1893, the Celticist Sir John Rhys found plenty of Manx speakers, but almost none of them children; there was little evidence of any desire of the speakers to preserve the language (Price 1984, 78–81).[63]

Manx quite obviously was moribund by the last quarter of the 19th century. This set the stage for its collapse between 1901 and 1921, when the speaking population fell from 4,657 to 896. From there it slid, almost inexorably, to extinction. The factors that induced moribundity (the establishment of English as the prestigious language, leading to the abandonment of the Manx for schooling and worship) were in place in the early 19th century; after actual moribundity set in by the 1870s or so, the language limped along until it crossed some invisible demographic threshold early in the last century, after which it rapidly declined to a terminal state.[64] So to say that Manx was alive up to the moment Ned Maddrell drew his final breath, and then suddenly it was not, is to ignore the fact that the language had been functionally dead as a mother tongue for most of his life. That is one of the tragedies of language extinctions. For speakers of a critically endangered language, its final decline and extinction unfold—sometimes insensibly, sometimes wrenchingly—as part of a complex transformation of their collective mental landscape, in which something once taken for granted as vigorous and enduring becomes just a memory.

Yet Manx lives on even after death as a "language of heritage" for a small band of enthusiasts. The Manx Gaelic Society (Yn Cheshaght Ghailckagh, founded in 1899) continues to promote Manx as the national language of the Isle of Man (which has semi-autonomous status within the United Kingdom) and to encourage literature in Manx.[65]

But the hard fact is that the only way Manx will ever again be the mother tongue of the island is through a prolonged and concerted effort by a majority of its residents. In today's political and socioeconomic circumstances, it is not easy to see how such an effort could be mustered. If Manx can be preserved as a supplemental language of heritage, good—yet it would be, by any linguistic or cognitive measure, a far cry from a robust mother tongue whose speakers are using it as an instrument to engage the entire range of everyday concerns and which itself is functionally evolving across a full range of social situations in response to the needs of its community of users. Languages of heritage will develop very differently than if they had maintained an unbroken existence as a mother tongue.

At least Manx *has* a fallback position. Most extinctions in the years to come will be far less well attested to than was the death of Manx. They will not have the benefit of anything like Yn Cheshaght Ghailckagh, and many will die[66] before anyone has a chance to study or record them in depth.

THE BIOCULTURAL PRESENCE

If we look at the last 10,000 years of history, it is obvious that cultural diversity evolved within a matrix of diverse biological riches. And, conversely, the natural environment has been remade by widely differentiated human action (Thomas 1956; Vitousek et al. 1986, 1997; Turner et al. 1993). Our evolutionary history tells us that cultural diversity is not merely a referent of the bio-logical diversity of the nonhuman world, or vice versa. Rather than speak of the evolution of nature and culture as if they were on parallel tracks, it would be more accurate to focus on "evolutionary reciprocity," for it is the give-and-take between the two realms of living difference that has given form and mean-ing to the world we know today.[67] This brings up an important though often overlooked corollary point. The diminution of biodiversity and of cultural diversity are usually portrayed as two coincidental trends which (it may be granted) share some or even most of the same ultimate causes, but whose con-sequences are largely separate. In truth, the two crises we have been reviewing in this chapter are converging to produce a wholesale loss in the evolutionary reciprocity between nature and culture. What is being destroyed is the funda-mental process that generated the conditions of life that we ("we" meaning *all* species) are at home in.

These conditions of life, now at peril, are invested in what I call the "bio-cultural presence." The biocultural presence is nothing less than the entire complement of biological and cultural diversity now existing, bestowed upon Earth by millions of years of evolution. Biologically, it has shaped the species *Homo sapiens*; culturally (and spiritually, if you like) it is the foundation of what it means to be human.[68] It recognizes that nature and culture are not co-extensive, but that significant parts *are* indissolubly mixed; and that these areas of overlap, too often sublimated when biodiversity and cultural diversity are treated separately, are at the core of what is at risk in the current converging extinction crises. So instead of continuing dualities, we can now rightly turn to a consideration of *biocultural* diversity.[69] The next chapter looks further at species and languages as a means of charting the biocultural presence, and how comparing them gives a glimpse of the pluralistic nature of reality.

No one can now predict with any great accuracy what the final toll on species and languages will be. We can predict, however—and with consider-able accuracy—what the consequences will be. Diversity has been with us since life began some 3.9 billion years ago. It has been the backdrop to our own species since the origin of the human mind some million to 100,000 years ago.[70] If we indeed suffer a devastating loss of variety in nature and culture over the next hundred years or so, the face of the planet will have been changed beyond all our present knowing. And so too—though less obviously, maybe even imperceptibly—will have been the essential humanity of our own species. For the face of humankind is many faces; our voice, many voices. Diversity *is* the human identity.

We can now circle back to the vignettes with which we began this chapter. The future of the biocultural presence goes to the heart of what human consciousness has been and will become. Making distinctions, needing distinctions, is the natural condition of all life on Earth. We cannot "know," and cannot intelligibly arrange "what is," without recourse to diversity. It is a fundamental prerequisite to ordering and valuing existence, so fundamental that it is mostly transparent to our thoughts.[71] As a frame of reference, diversity permeates our lives while remaining, for the most part, beyond the edge of our attention—like plants disappearing from a ridge, or the plight of a sick peasant girl, off in some distant country.

That is how the loss of difference is being felt, from the global down to the personal level. Species and languages are evidently much more than mere arbitrary categories. Certainly they carry emotional significance, but what's more they apparently have at least some measure of objective reality. What does that mean? To find out, let us turn to an in-depth comparison of the similarities (and fundamental differences) between species and language. Once again we start by reaching backward in time; here, to review the parallels between how evolutionary biology and comparative linguistics work. Along the way some unexpected discoveries await that show how comparing species and languages is far more than a metaphorical exercise. Once that point is established, we will explore a radically different approach to forming categories. The reward is a theoretical understanding of why species and languages are the keys to biocultural diversity.

3

SPECIES, LANGUAGES, AND THE STRUCTURE OF DIVERSITY

On the face of things, it is remarkable that two fields of inquiry so seemingly different as biology and linguistics should have followed such similar paths toward the same fundamental conclusion: that the process of change is evolutionary, with its course to the present traceable to ancestors. In both nature and language, the pathways reaching back in time form coherent lineages that contain some of the deepest truths of history (see Mishler 2001, 78). These tenets, though second nature to us now, shattered the intellectual world of Western Europe in the nineteenth century. They displaced an insular, dogmatic outlook in which many baffling questions—such as the reason for the great plenitude of languages and species on Earth—could be explained only by convoluted reference to a largely static, largely predetermined grand order represented by such conceptions as the Great Chain of Being. As we saw in Chapter 1, that type of thinking ran out of steam at the end of the 18th century. Then, thinkers in a number of fields, including linguistics, began groping toward an embryonic conception of evolution (Ruhlen 1991, 25–28).

This was all long before Darwin, of course. His great achievements in *On the Origin of Species* (published in 1859) were to crystallize and exemplify the concept of evolution and then elaborate it with the theory of natural selection. In so doing, he managed to consolidate the strains of evolutionary thought then current in such fields as geology, paleontology, and biology itself. Historical linguistics, it must be said, needed no such consolidation, as its basic comparative methods were already well established by the work of Rasmus Rask, Franz Bopp, Jacob Grimm, and others working in the early 1800s (Jespersen 1922, 22 ff.; Pedersen 1962, 248–262). All these men were following methods that had been signalized by Sir William Jones in his famous Asiatick Society address of 1786, in which he was the first to propound the common ancestry of Sanskrit and Greek, thus laying the foundations of Indo-European philology. Nonetheless, the astounding speed with which Darwin's views on species gained ascendancy set something of an imprimatur on what had been happening in linguistics.

In fact, Darwin himself felt a strong affinity between his own field and historical linguistics. In the *Origin* he ventured the opinion that, if we could but properly classify all the languages of the world, both living and extinct, along with "all intermediate and slowly changing dialects," we would of necessity arrive at "a perfect pedigree of mankind" (Darwin 1859, 422–423). He soon found this line of reasoning confirmed and extended by one of his mentors, the geologist Charles Lyell, for whom Darwin held great respect. "No praise can be too strong," he declared, "for the inimitable chapter on language in comparison with species" in Lyell's 1863 book *The Antiquity of Man* (Darwin 1892, 270).[1] What so impressed Darwin was the chapter titled "Origin and Development of Language and Species Compared." In it, Lyell took the state of the art in historical linguistics[2] and drew parallels with species based on the ideas charted in the *Origin*.

Lyell's chapter is well worth reading even now, for he discusses in fairly modern terms some linguistic problems that are still very much with us, such as the time-depth at which various language families branched off, the relative speed of language change, and the difficulty of distinguishing languages from dialects. What struck Darwin as "most ingenious and interesting" (Darwin 1892, 268) was Lyell's conclusion that "all the existing languages, instead of being primordial creations, or the direct gifts of a supernatural Power, have been slowly elaborated," just as Darwin had claimed for species in the *Origin*. This is true even though (like gaps in the fossil sequence) there are "no memorials of all the intermediate dialects" that must have existed between the time proto-languages were spoken and the present (Lyell 1863, 458). The lack of such linguistic "fossils" is not surprising, Lyell maintained, because "it is not part of the plan of any people to preserve memorials of their forms of speech expressly for the edification of posterity,"[3] most writing being purely practical and thus susceptible to being discarded as the years go by. Yet there is proof of antecedent forms of language in those written records that do exist: absent special study, Lyell's fellow English speakers could not understand the Anglo-Saxon of King Alfred's day, yet the two are obviously related. A further proof of language evolution is the fact that languages maintain geographic relations: "modern Romance languages are spoken exactly where the ancient Romans once lived and ruled," as is modern Greek, and so on (Lyell 1863, 459, 461). If languages were primordial, unchanging, special creations—as species were commonly held to be prior to Darwin—then each one would be an isolate: we would find languages as different as Basque, Korean, and Zuni, say, all jumbled together.

Darwin had revealed the true character of species, and for Lyell, the parallels with language were clear: both are "liable to slow or sudden extinction. They may die out very gradually in consequence of transmutation, or abruptly by the extermination of the last surviving representatives of the unaltered type" (Lyell 1863, 466). With each new generation of speakers, a language's vocabulary is subjected to two opposing tendencies: persistence and innovation. In the organic world, these correspond respectively to "the force of inheritance" and "the variety-making power of animate creation" (Lyell 1863, 467). And selection works in language as well as in nature:

> The slightest advantage attached to some new mode of pronouncing or spelling, from considerations of brevity or euphony, may turn the scale, or more powerful causes of selection may decide which of two or more rivals shall triumph and which succumb. Among these are fashion, or the influence of an aristocracy, whether of birth or of education, popular writers, orators, preachers,—a centralized government organising its schools expressly to promote uniformity of diction, and to get the better of provincialisms and local dialects. Between these dialects, which may be regarded as so many "incipient languages," the competition is always keenest when they are most nearly allied, and the extinction of any one of them destroys some of the links by which a dominant tongue may have been previously connected with some other widely distinct one. It is by the perpetual loss of such intermediate forms of speech that the great dissimilarity of the languages which survive is brought about. (Lyell 1863, 463–464)[4]

Darwin picked up on Lyell's enthusiasm and published similar, if less lengthy, comparisons between species and languages in *The Descent of Man*.[5]

The relevance of the Darwinian revolution was grasped immediately by such linguists as August Schleicher. Four years after the publication of the *Origin*, Schleicher came out with a pamphlet called *Darwinism Tested by the Science of Language*. "What Darwin now maintains with regard to the variation of the species . . . has been long and generally recognised in its application to the organisms of speech," he observed dryly. "To trace the development of new forms from anterior ones is much easier, and can be executed on a larger scale, in the field of speech than in the organisms of plants and animals. . . . The kinship of the different languages may consequently serve . . . as a paradigmatic illustration of the origin of species . . ." (quoted in Law 1990, 819).

Like Darwin, Schleicher thought languages hold the key to the pedigree of humankind. He claimed that the classification of people into races should not be based on physical appearance or any other external criteria, "as they are by no means constant, but rather on language, because this is a thoroughly constant criterion. This alone would give a perfectly natural classification system." The linguist Otto Jespersen, who provides this paraphrase (1922, 75), goes on to demolish the idea, asking whether we are "to reckon the Basque's son, who speaks nothing but French (or Spanish), as belonging to a different race from his father?" (1922, 75).

Indeed, a counter-reaction against taking the language–species parallel too far had set in very shortly after Schleicher's death in 1868. Among the things the "Young Grammarians" (*Junggrammatische Richtung*, the generation of German linguists who began to come of age in the 1870s) revolted against was Schleicher's assertion that each language is "a natural organism and its study a natural science" (Collinge 1990, 880–881). So did linguists outside Germany, such as William Dwight Whitney, who wrote in 1875 that "physical science on the one side, and psychology on the other, are striving to take possession of linguistic science, which in truth belongs to neither" (quoted in Law 1990, 820). Vivien Law goes on to say that "this plea for the autonomy of linguistics was to echo on through the twentieth century" (Law 1990, 820).[6] She is quite right. Here is Jespersen again, on the opening page of *Language: Its Nature, Development, and Origin*, published in 1922:

> The distinctive feature of the science of language as conceived nowadays is its historical character: a language or a word is no longer taken as something given once

for all, but as a result of previous development and at the same time as the start-
ing-point for subsequent development. This manner of viewing languages constitutes a decisive improvement on the way in which languages were dealt with in previous centuries, and it suffices to mention such words as 'evolution' and 'Darwinism'
to show that linguistic research has in this respect been in full accordance with tendencies observed in many other branches of scientific work during the last hundred
years. Still, it cannot be said that students of language have always and to the fullest
extent made it clear to themselves what is the real essence of a language. Too often
expressions are used which are nothing but metaphors that obscure the real facts
of the matter. Language is frequently spoken of as a 'living organism'; we hear of
the 'life' of languages, of the 'birth' of languages and of the 'death' of old languages,
and the implication, though not always realized, is that a language is a living thing,
something analogous to an animal or a plant. Yet a language evidently has no separate existence in the same way a dog or a beech has, but is nothing but a function
of certain living human beings. Language is activity, purposeful activity, and we
should never lose sight of the speaking individuals and of their purpose in acting
in this particular way. (Jespersen 1922, 7)[7]

N. E. Collinge, reviewing all this, remarks, contrary to Schleicher and the
other "linguistic Darwinists of the last century," that "languages and their
forms are not some kind of natural organism" (Collinge 1990, 899). This
conclusion, while valid enough, is really something of a straw man. No biologist would ever equate a language with an individual member of a species;
the two aren't units at the same level of comparison, for the simple reason that
humans can choose which language or languages to use,[8] whereas volition does
not play a role in determining which species an organism belongs to. Try as it
might, a toucan can never become a jaguar—though the whole point is that
it is quite beyond the toucan's capacity to try. No nonhuman species can consciously direct its own collective future, while the course of a given language
is, ultimately, a matter of choices made by both speakers and nonspeakers. To
put it another way, languages change by cultural selection; species, by natural
selection.[9]

A more pointed criticism comes from Johanna Nichols in her *Linguistic Diversity in Space and Time*. There, reviewing different senses of the term "evolution"
as they apply to linguistics, she says: "No evidence of anything like speciation
has been found in this or any other typological work. Although linguistics has
no analog to the biological notion of species, it is safe to say, informally speak-

ing, that languages and linguistic lineages are related to each other as individuals or kin groups of a biological species are, not as species in a genus" (Nichols
1992, 277). Again, one must at least partly agree with her, for undoubtedly no
language has ever diverged from all others to the point that it is absolutely
impenetrable, is not capable of being learned by outsiders, is—to put it in biological terms—"reproductively isolated" from other languages. Such an occurrence would indeed make for an entirely different "species" of language in the
strictest sense of the word. But I strongly disagree that there is no linguistic
analogue to biological species; to my mind, an analogy is a comparison in which
the broad outlines run parallel while the particulars may not. There is a convincing analogy to be drawn between a species and a language. More importantly, the analogy is far from being one of Jespersen's "metaphors that obscure
the real facts of the matter"; rather, it is useful to a practical consideration of
the nature and extent of linguistic diversity.

SIMILARITIES BETWEEN SPECIES AND LANGUAGES

So a language is not the same kind of a thing as a species. Yet languages and
species *are* similar in many powerful ways. This should not be terribly surprising, since both are, in the most general terms, expressions of cohesion for
groups of individuals. They are givers of identity, bestowers of distinctiveness.
That broad, abstract similarity is valuable in itself because it gives us a basis on
which to compare diversity in language and nature. But there are also more
concrete similarities, beginning with the problem of how to define species and
languages themselves.

The puzzle of how to distinguish languages from dialects, and indeed languages from other closely related languages, is of long standing.[10] To many linguists, it is well-nigh intractable. Martin Durrell, for example, is not alone when
he asserts that there are "no satisfactory objective criteria" for determining
what is a language and what a dialect. As he rightly points out, the layperson's
criterion, mutual intelligibility, is by itself not definitive, because there are many
mutually intelligible autonomous languages and, conversely, mutually unin-
telligible dialects of the same language. "In fact, in the conventional designa-
tion of a particular geographically determined variety of a 'dialect' of a particular 'language,' our definition of 'language' is still based on geopolitical,
ethnic or cultural rather than linguistic criteria . . ." (Durrell 1990, 921). Furthermore, as Tove Skutnabb-Kangas has shown, none of the other criteria that

have been proposed (structural similarity, level of understanding, existence of standardization) work either. Sharpening Durrell's formulation, she sees the crux of the issue as the relative political power of the speakers being classified (Skutnabb-Kangas 2000, 15).[11] Thus the common saying among linguists that "a language is a dialect with an army."

Vexed by the difficulty of distinguishing languages from dialects, of separating discrete from continuous variation in language, it is easy for linguists to look at biology as being far more precise than their own field:

> The language/dialect distinction in linguistics parallels the species/variety distinction in biology. Whereas a species is usually thought of as a group of plants or animals that can interbreed and produce viable progeny, a language is a group of dialects among which there is mutual intelligibility. . . . But the analogy is not perfect, since the criterion of mutual intelligibility is less well defined than that of interbreeding. Mutual intelligibility ranges from close to 0 percent to close to 100 percent; interbreeding is usually an all-or-nothing proposition. . . . [T]here are many gray areas in deciding what to call a language and what to call a dialect. Biological speciation is much more clear-cut. (Ruhlen 1991, 21)

In truth, this conclusion by the linguist Merritt Ruhlen is much too generous to biology. Although there is a strong case to be made for the mode of speciation alluded to above, the issue is far from settled. Proponents of the biological species concept (of which more below) argue that the predominant (or even the only) way species form is by "allopatric speciation," in which a population becomes geographically isolated from the rest of the species so that it eventually becomes reproductively isolated as well and can no longer interbreed (Mayr 1963, 451–480). However, there is a strong undercurrent of dissent to this viewpoint, centered among certain botanists, genetic biologists, and taxonomists. They point out that there are plenty of cases of "sympatric speciation" in which prolonged geographical isolation is not a prerequisite for the formation of a new species.[12] Edward O. Wilson comes down squarely in favor of the primacy of allopatric speciation, but acknowledges that "it is supplemented in nature by a rich medley of other modes of speciation" (Wilson 1992, 69; see also Wilson 2000).

Speciation is not, therefore, quite so clear-cut. In fact, biologists have just as much trouble defining species as linguists do language.[13] Surprising as it may be to many, the very validity of the species concept is still up for debate:

What is a species? This simple question has troubled biologists for more than two centuries. Although accepted so widely as a 'natural,' basic or fundamental unit, many conflicting definitions of species have been coined, and agreement is still lacking. The range of definitions reflects, to a large degree, the differing interests and differing theories of individual scientists about the origin of diversity itself—literally from Genesis to Darwin and DNA. (Groombridge 1992, 13)

There are two main competing ideas about what constitutes a species. One, called the "biological species concept," is what most of us think of when we think of a species. "The basic idea of a biological species is that of a 'pool' of genes available for re-combination through sexual reproduction, but not with genes belonging to other gene pools, from which they are 'protected' by a variety of recognition and isolation mechanisms (behavioural, physiological, genetical, etc.). Thus the biological species to which a given individual belongs is determined by the limit of the populations with which it interbreeds, or potentially interbreeds." This concept is "probably the most widely accepted view of the species held by biologists today" (Groombridge 1992, 14).[14]

It is not beyond controversy, however. Many others argue forcefully that the biological species concept, while intuitively attractive, is simply inapplicable to great numbers of organisms (Cracraft 1989; Templeton 1989).[15] This camp makes two basic objections: first, that the concept simply ignores the large class of organisms that reproduce asexually or self-mate; second, that there are so many exceptions to the concept, even among sexually reproducing organisms, that it loses much of its validity. These exceptions involve evidence that there is considerably more gene flow *between* species than should take place if they were truly reproductively isolated, as the biological species concept holds. In other words, on the "fringes" of numerous species there is a lot of genetic exchange going on (see MacArthur 1972, 71–73)—very much like the linguistic exchange that goes on between dialects along the margin of their language's ranges. Such naturally occurring hybridization is commonplace among plants (botanists have a special term, the "syngameon," to denote aggregations of plant species that exhibit gene exchange); now, with the advent of genetic fingerprinting techniques, there is evidence that it takes place even among mammals (Templeton 1989, 11; Templeton 1991). This violates the most basic tenet of the biological species concept: the separateness of gene pools (Groombridge 1992, 14).

Those who criticize the biological species concept have proposed alternatives that either emphasize (1) mechanisms of reproductive cohesion rather than isolation (Templeton 1989, 12–24) or (2) differentiation signified by historical relationships between whole taxa rather than current blood relationships of individuals. The latter alternative, standing in opposition to the biological species concept, is known as the "phylogenetic species concept." It views species as irreducible clusters of organisms "diagnosably distinct from other such clusters, and within which there is a parental pattern of ancestry and descent" (Cracraft 1989, 34–35).[16] The problem with the phylogenetic species concept, according to its critics, is that in the search for "diagnosable distinctions" one could lose sight of the forest for the trees and, "by reductio ad absurdum, every population, stage, morph,[17] or even individual organism could be elevated to separate species status" (Groombridge 1992, 15). And this debate is just the tip of the iceberg, for dozens of species concepts have been forwarded: everything from "recognition species" to "chimaeric species" (Harrison 1998, 20–21; Corliss 2000, 136).[18]

SPECIATION AND LANGUAGE GENESIS

It is not to our purpose to go into this debate in any further detail. What has been said so far points to four important things to keep in mind when comparing species and languages, and the larger process of speciation with language genesis. First, those who wish to define species must face exactly the same kind of problem as those who would define languages: that of separating discrete from continuous variation at the margins, where species blend into species and languages blend into languages (or dialects into dialects). It is striking that the two defining criteria at issue—reproductive isolation for species, and mutual intelligibility for languages and dialects—are both intuitively appealing, and work quite well in many instances, yet are both faulty in enough instances so that neither is alone sufficient to make a clean definition. This is an important point to which we will return.

The second point is that "the explanation of geographic and temporal variation in species diversity is one of the central problems of biology. It has also proved to be one of the most intractable. The problem has generated an enormous amount of literature in which many different hypotheses have been proposed to attempt to account for it; *these hypotheses often operate at different levels of explanation* and much confusion has arisen as a result" (Groombridge

1992, 46; emphasis added). The same can be said for discussions of linguistic diversity. This problem of differing levels of explanation is examined below.

The third point to keep in mind is that speciation and language genesis occupy similar positions of importance with respect to the creation and continuation of diversity. "Speciation is potentially a process of evolutionary rejuvenation, an escape from too rigid a system of genetic homeostasis. . . . The importance of speciation is that it invites evolutionary experimentation. It creates new units of evolution, particularly those that are important for potential macroevolution. Speciation is a progressive, not a retrogressive, process" (Mayr 1963, 555). So is language genesis with respect to cultural diversity. Culture requires both continuity *and* vitality if it is to have meaning, and freely evolving mother-tongue languages are the major vehicle of both. They are the building blocks of cultural diversity, arguably the fundamental "raw material" of human thought and creativity.[19] If their numbers are reduced dramatically, then the raw material for human creative evolution is diminished, eventually making the world's cultures increasingly monolithic, with the range of cultural variety severely circumscribed.

The fourth point is that the idea of allopatric speciation being supplemented by other modes is mirrored in the processes of language genesis. As with speciation, a primary mechanism of language genesis is isolation, which follows geographic separation—in this case, communicative rather than reproductive isolation. Historically, when a language community split into two or more groups that then remained out of contact with each other, forms of speech once mutually intelligible gradually drifted apart, "provided only that they were isolated from each other for a sufficient period of time"—roughly 500–1,000 years, according to Ruhlen (1991, 6). To borrow the biological term, what is described here is "allopatric language genesis." It is apparent that geographic isolation has often played a key role in allowing the rise of differentiated languages. "Of all the kinds of variation in and among languages," declared Morris Swadesh, "geographic variation is probably the most universal. Social differences in language are very important in some places and less so in others, but geographic variation is found everywhere. Furthermore, as we go back in time some thousands of years, the nongeographic factors are probably everywhere far less important" (Swadesh 1971, 18). Indeed, this is the premise undergirding the human origination–migration debate, in which linguistics has become enmeshed.

In biology, "geographic isolation is a purely extrinsic and completely reversible factor which does not by itself lead to the formation of species. Its role is simply to permit the undisturbed genetic reconstruction of populations that is the prerequisite for the building up of [reproductive] isolating mechanisms" (Mayr 1963, 554). This principle seems to hold true for language genesis as well. If two speech communities of the same language become geographically separated and communicatively isolated for a period, but decide to reunite before total divergence sets in, then presumably whatever differences have arisen during the interim will sooner or later be reconciled and the language will not split. Thus we see a three-step process in this allopatric mode of both speciation and language genesis: (1) migration or separation that results in geographical isolation for a long period, so that there develops (2) genetic/speech adaptations that result in reproductive/communicative isolation from the parent population, ultimately producing (3) differentiation (genetic or linguistic) by the group as a new species or language.

However, as in biology, the allopatric mode doesn't explain all language distributions and the linguistic variation within them. It is clear that many distinct languages and dialects have developed in close quarters, that is, in situations where the nexus of geographic and communicative isolation does not exist. There are at least two ways in which such "sympatric language genesis" (as we might call it) can take place. One is through externally imposed social isolation of a particular group, as when class differences become marked by divergent linguistic forms through various means of social and economic pressure. A second is through self-imposed social boundary marking, as when a group of people consciously seek to distinguish themselves by their speech. Such boundary marking may be antagonistic in its origins, but need not continue so once group distinctiveness has been established. For example, the linguistic diversity of Aboriginal Australia seems at least partly due to mutually enriching intercultural communication among adjacent communities. This positive form of sympatry requires an equilibrium among relatively stable societies (Mühlhäusler 1995).

The linguist R. M. W. Dixon has proposed a historical model of language change in which these equilibria have been occasionally "punctuated" (i.e., sharply disrupted) by rapid changes in social and economic conditions (Dixon 1997).[20] Examples include epidemics, famines, and the introduction of innovative weaponry or some other far-reaching technology. Throughout most of history, according to Dixon, equilibrium has returned after these episodic dis-

ruptions. However, the last few hundred years "have been the scene of a spectacular punctuation," on a scale without precedent, marked by the "rise of
world religions, imperialism, guns, writing, and other factors" constituting
what we think of as modernity. They "have combined in such a way that some
people and their languages have grown more powerful, and swept all before
them" (Dixon 1997, 4).

Another linguist, Nancy Dorian (1998, drawing on the work of Ernest Gellner and Ralph D. Grillo), gives us an example of the beginnings of standard-
language dominance from France at the time of the Revolution. French society
at the dawn of the industrial age was an "agro-literate polity" (Gellner's term)
in which the most prestigious social strata—the nobility, clergy, and mer-
chants—overspread the polity as a whole. Meanwhile, the lower socioeconomic
levels of society were characterized by "a variety of distinct small communities," many of which did not speak "standard" French, but rather markedly distinctive regional varieties (such as Alsatian or Provençal) or completely unrelated languages (such as Breton or Basque). In societies organized like this, the
role of the state was pretty much limited to levying taxes and keeping the
peace; it had no interest in promoting intercommunication among its sub-
communities. Indeed, "linguistic heterogeneity had been useful to the crown
as a means of keeping various feudal constituencies from making common
cause with one another."

In industrial societies, by contrast, conditions are quite different. Industrial
means of production require universal literacy and numerical skills such that
individuals can communicate immediately and effectively with people previously unknown to them. Forms of communication must therefore be stan-
dardized and able to operate free of local or personal context. This in turn
places great importance on educational capacities that allow people to be slot-
ted and re-slotted into a variety of economic roles. The state is the only orga-
nizational level at which an educational infrastructure of the necessary size and
costliness can be mounted (Dorian 1988, 6; Gellner 1983, 35–36).

The passing of agro-literate society brought "a new focus on the polity as a
totality" so that, rather suddenly, "the fact that a number of sizable subcom-
munities such as the Bretons, Basques, Alsatians, and Occitanians were inca-
pable of understanding and speaking French became unacceptable." French
would have to become the language of Revolutionary France (Dorian 1998,
6–7; citing Grillo 1989, 30, 34, 35). So it was that the 1790s saw the Abbé Henri
Gregoire calling for the universalization of French and the Marquis de Con-

dorcet propounding a Universal State with French as a leading candidate for the lingua franca (Todorov 1993, 23–25; Dorian 1998, 6).

To generalize this picture, we can say that the preindustrial equilibrium periods of history were characterized by ubiquitous agro-literate polities: societies were premised upon people knowing a great deal about their immediate environment and the agricultural (or subsistence) uses that could be made of it. Stated in current terms, localized knowledge of agrobiodiversity was the common currency of every community, large or small, indigenous or not. Even city dwellers and the nobility were only distanced, not divorced, from their local nature-based knowledge system. Before the Industrial Revolution, disruptions to the equilibria were episodic and not socially pervasive, and so did not result in wholesale language shifts because the underlying agro-literate polities remained intact.

Industrialization changed this forever, and that is why it is such a decisive break in history. Of great significance is the fact that, in industrialized society, communication must be able to "operate free of local or personal context." This is precisely the opposite of what is required in preindustrial and indigenous civilizations, where the local and personal context is not merely paramount, but constitutes the entire organizing principle. With only a few exceptions (e.g., Latin in the period of the Roman Empire), throughout history languages evolved and diversified to fulfill local and personal contexts. Only after the Industrial Revolution—which, it should be emphasized, didn't pop up out of nowhere, but was seeded in the beginnings of European exploration and conquest—have several "world languages" been able to take hold simultaneously.

The parallels with biological species are obvious. Most species are local or regional in extent, and evolved over long periods during which they were subjected to merely episodic negative disturbances by relatively small numbers of humans. Just as the habitats of native flora and fauna can be threatened by invasive species and outright conversion, languages are threatened by sustained disruptions to the cultural, social, and economic settings in which they have evolved. What happens, as the linguist Stephen Wurm describes it, is that these traditional settings are abruptly "replaced by new and quite different ones as a result of irresistible culture contact and clash, with the traditional language unsuited for readily functioning as a vehicle of expression of the new culture" (Wurm 1991, 2–3).[21] And, just as modern transport allows invasive species to travel all over the world, global telecommunications means that economically dominant languages are now heard daily in the remotest regions. Geographic isolation no longer results in communicative isolation. Sympatric language genesis remains as a possibility, but even in geographically isolated places the stable social equilibria needed for mutual cultural enrichment are threatened with disruption. In short, the continuation of large-scale speciation and language genesis is now threatened by the elimination of the conditions that historically made them possible.

ENDEMISM IN SPECIES AND LANGUAGES

Species and languages are not just comparable on an abstract, conceptual level. There is also a pronounced pattern of congruity in the geographical distribution of the two. What is especially striking is that many countries with high numbers of endemic species also have high numbers of endemic languages. *Endemic species* are those found in restricted locales and nowhere else. In compilations of global data, endemism usually is pegged to individual countries (as in Groombridge 1992, 137, 155). Endemic species are important because they represent unique adaptations to environmental conditions in a relatively small area. For this very reason they are vulnerable to predation or competition by less-specialized invasive exotic species. The concept can be readily extended to languages, with *endemic languages* being those restricted to a single country.[22] Endemic languages no doubt hold a high percentage of the unique traits in human language. This is borne out by analysis of the twelfth edition of *Ethnologue* (Grimes 1992b), which reveals that just over 83% of the world's languages (5,635 out of 6,760) are endemics.[23] These languages are also vulnerable to "predation and competition" from dominating and invasive languages that are larger and more politically powerful.

Of the 25 countries with the highest number of endemic higher vertebrate species (i.e., mammals, reptiles, and amphibians), 16 are also among the top 25 in endemic languages—a concurrence of 64% (Table 3.1; Figure 3.1 maps the pattern). It seems highly unlikely that this is mere coincidence. There are several geographical and environmental factors that explain why so many countries appear on both lists:

- Extensive countries with highly varied terrain, climate, and ecosystems tend to have many endemic species simply because of their size and biophysical diversity. These same factors, operating at the lower population levels that prevailed before European expansion, also fostered commu-

Table 3.1

Endemism in Language and Higher Vertebrates: Comparison of the Top 25 Countries

Endemic Languages (Number)	Endemic Higher Vertebrate Species (Number)
1. **Papua New Guinea** (847)	**Australia** (1,346)
2. **Indonesia** (655)	**Mexico** (761)
3. Nigeria (376)	**Brazil** (725)
4. **India** (309)	**Indonesia** (673)
5. **Australia** (261)	Madagascar (537)
6. **Mexico** (230)	**Philippines** (437)
7. **Cameroon** (201)	**India** (373)
8. **Brazil** (185)	**Peru** (332)
9. **Democratic Republic of Congo** (158)	**Colombia** (330)
10. **Philippines** (153)	Ecuador (294)
11. **United States** (143)	**United States** (284)
12. Vanuatu (105)	**China** (256)
13. **Tanzania** (101)	**Papua New Guinea** (203)
14. Sudan (97)	Venezuela (186)
15. Malaysia (92)	Argentina (168)
16. **Ethiopia** (90)	Cuba (152)
17. **China** (77)	South Africa (146)
18. **Peru** (75)	**Democratic Republic of Congo** (134)
19. Chad (74)	Sri Lanka (126)
20. Russia (71)	New Zealand (120)
21. **Solomon Islands** (69)	**Tanzania** (113)
22. Nepal (68)	Japan (112)
23. **Colombia** (55)	**Cameroon** (105)
24. Côte d'Ivoire (51)	**Solomon Islands** (101)
25. Canada (47)	**Ethiopia** (88); Somalia (88)

Countries appearing on both lists are in bold. Figures for Ethiopia include Eritrea. Higher vertebrates include mammals, birds, reptiles, and amphibians; reptiles not included for United States, China, and Papua New Guinea because the number of endemic reptile species is not reported in the source table.
Source: Harmon 1996, 97; using figures derived from Grimes 1992b (languages) and Groombridge 1992, 139–141 (species).

nicative isolation among small speech communities, thus allowing many small autonomous languages to evolve. Examples from Table 3.1 include Mexico, the United States, Brazil, India, and China.

- Island countries tend to have high numbers (and often a high density) of endemics because their physical isolation from continental land masses has allowed more locally adapted species to develop. Those that in addition have broken terrain (i.e., a variety of elevational gradients) or some other physical barriers to seed or animal dispersal are also predisposed to higher species endemism (Heywood 1995, 180). The same conditions also tend to produce more endemic languages on these islands than in similar continental countries. Examples from Table 3.1 include Papua New Guinea,[24] Philippines, and Solomon Islands.

- Tropical countries tend to have more species than those with predominantly boreal, temperate, Mediterranean, or austral (Southern hemisphere) climates (Groombridge 1992, 43). Tropical countries also tend to have more endemic languages than the others because the warm climate and ample rainfall, coupled with the richness of the natural resources at hand, was historically favorable to the small hunter–gatherer societies that flourished before the advent of concentrated agriculture. With these small, mobile cultural groups came many endemic languages. Examples from Table 3.1 include Cameroon, the Democratic Republic of Congo (formerly Zaïre), and Tanzania.

Where all three of these conditions coincide within a country, there comes the possibility of extremely high endemic richness in both species and languages. Indonesia, which ranks fourth on the vertebrate species and second on the language list, is a textbook example of this. Australia, which ranks in the top five of both lists, combines some of the factors.

Linguistic diversity is, of course, a product of social as well as geographical and environmental factors, and a full accounting of the genesis and spread of languages must reflect the entire picture. In his 1999 study *Linguistic Diversity*, Daniel Nettle emphasizes how local or regional ecological conditions pose unique environmental problems for the humans living in those particular places, how differing social networks are shaped by the need to solve these problems, and how linguistic diversity arises out of this process of environmental–social interaction. Examples of the challenges posed by different environments include droughts and floods, temperature extremes, unpredictable crop-killing freezes, and fluctuations in populations of game animals that are hunted for subsistence. Generalizing, Nettle subsumes them under the term "ecological risk": the probability that a household will face a temporary shortfall in food production. Nettle sees ecological risk as a key determinant in the

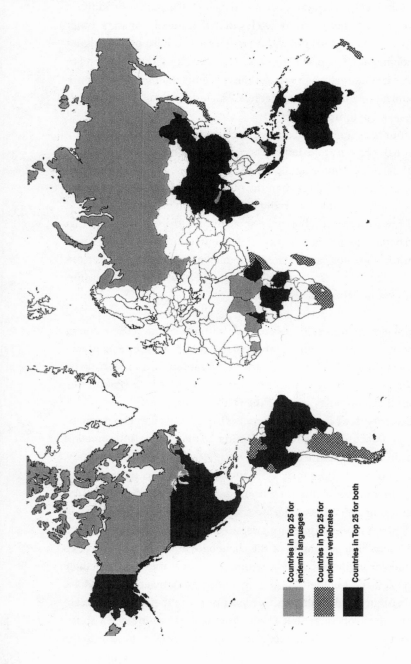

Figure 3.1. Endemism in language and higher vertebrates: Comparison of the top 25 countries. Derived from data in Harmon 1995 and Harmon 1996.

Countries in Top 25 for
endemic languages

Countries in Top 25 for
endemic vertebrates

Countries in Top 25 for both

geographical distribution of languages (Nettle 1999a, 79). He measures ecological risk using climatological data to calculate a country-by-country mean
growing season, and then statistically associates those data with each country's
linguistic diversity (using language richness, both endemic and nonendemic, as
a proxy). Nettle hypothesized that the greater the ecological risk, the fewer the
languages that will be found in a country of a given size and population. His
analysis (which, however, excludes nontropical and certain tropical countries)[25]
generally supports the hypothesis (Nettle 1999a, 82–96). Not surprisingly, this
is in accordance with the above discussion of geographical and environmental factors involved in producing language endemism, for, given that endemic
languages make up such a high proportion of the overall total, their distribution may be taken to stand for the distribution of languages as a whole.

If that is so, then one might expect the distribution of endemic languages to
match up well with that of species as a whole, *both* endemics and nonendemics.
There is evidence that this is exactly the case. Nearly every region rich in species
is also rich in endemic languages. Central Africa, Amazonia, southeastern Asia,
Australia, and China conform to the pattern. The same is true on a national
basis: The list of countries with the most endemic languages corresponds very
well with a list of the "megadiversity countries" identified by Conservation
International as having a large fraction of the world's species diversity in
selected groups of animals and plants. Table 3.2 shows that 13 of the 17
megadiversity countries (76%) are also among the top 25 in endemic languages.
Finally, the pattern carries over to a comparison of endemic languages with
flowering plant species, where the concurrence between the top-25 lists is 17
of 25, or 68%, and with endemic bird areas as identified by BirdLife International, where the concurrence is 12 of 19, or 63% (Table 3.2).[26]

The decline of the conditions that enabled biocultural endemism to develop
is a symptom of the underlying loss of evolutionary reciprocity discussed in
Chapter 2. The burden of that loss falls heavily on indigenous peoples whose
societies have been built around an intimate association with their immediate
environment. Traditional indigenous environmental knowledge, so different
in form and content from dualistic Western science, has almost completely
internalized the evolutionary give-and-take between nature and culture. The
irony is that traditional environmental knowledge is disappearing just as it is
starting to be accepted as valid scientific information by some Western scientists (e.g., Wilson 1992, 42–44).[27] As more and more indigenous youth are
absorbed into the swelling tide of globalized pop culture, their capacity to

Table 3.2

Endemism in Language Compared with Rankings of Biodiversity

Country	Rank Endemic Languages	Endemic Vertebrates	Flowering Plants	Endemic Bird Areas (EBAs)	On Mega-diversity List?
Papua New Guinea	1	13	18	6	Yes
Indonesia	2	4	7	1	Yes
Nigeria	3				No
India	4	7	12	11	Yes
Australia	5	1	11	9	Yes
Mexico	6	2	4	2	Yes
Cameroon	7	23	24		No
Brazil	8	3	1	4	Yes
Democratic Republic of Congo	9	18	17		Yes
Philippines	10	6	25	11	Yes
USA	11	11	9	15	Yes
Vanuatu	12				No
Tanzania	13	21	19	14	No
Sudan	14				No
Malaysia	15		14		Yes
Ethiopia	16	25			No
China	17	12	3	6	Yes
Peru	18	8	13	3	Yes
Chad	19				No
Russia	20		6		No
Solomon Islands	21	24			No
Nepal	22		22		No
Colombia	23	9	2	5	Yes
Côte d'Ivoire	24				No
Canada	25				No

Flowering plants include both endemics and nonendemic species. Twelve of the top 19 countries for endemic bird areas are also in the top 25 for endemic languages. "Megadiversity countries" are those identified as likely to contain a large percentage of the global species richness. Thirteen of the 17 megadiversity countries are also in the top 25 for endemic languages; the 4 that are missing are South Africa, Venezuela, Ecuador, and Madagascar.

Source: Modified from Harmon 1998b, 8; using data from Harmon 1995, 22–28 (endemic languages); Groombridge 1992, 139–141 (endemic vertebrates) and 80-83 (flowering plants); Stattersfield et al. 1998, 38 (endemic bird areas; only the top 19 countries are given); Mittermeier, Mittermeier, and Gil 1997 (megadiversity countries).

absorb the place-based knowledge of their ancestors is diminishing. Reports such as those chronicling the decay of the biological lexicon of the Tohono O'odham in Arizona (Nabhan and St. Antoine 1993; Hill 2001) are therefore distressing from both a biological and linguistic perspective.[28]

REASONS FOR DECLINE

Many linguists have noted similarities between the critical situation facing the world's biological and linguistic diversity (e.g., Zepeda and Hill 1991; Hale 1992; Krauss 1992; Maffi 1998; Skutnabb-Kangas 2000; Maffi 2001a). The main reasons for the declines, according to the linguist Stephen Wurm, can be characterized as "changes in ecology." A species can lose its viability through destruction of its habitat or because of the "introduction of other animal or plant species which in some important respects are more powerful and with which the species concerned is unable to compete successfully." For languages, changes in the social environment can mean that "the cultural and social settings in which a given language had been functioning, usually for a very long time, have been replaced by new and quite different ones. . . . The newly introduced dangerous animal and plant species mentioned above can be compared with negative and destructive attitudes towards this traditional language by the carriers of the newly introduced culture and speakers of the language serving as its means of expression" (Wurm 1991, 2–3).

The analogy between the destruction of natural habitat (for species) and the traditional social setting (for languages) is right on the mark. Species and languages have evolved over hundreds or thousands of years to adapt to very specific contexts (hence the frequent occurrence of endemism). If those contexts undergo unprecedented rapid change—as the world's environments and cultures are now experiencing—many species and languages will likely lack the resiliency (or the time) to adapt to the new conditions. As suggested above, the toll will be disproportionately borne by endemic species; for instance, 91% of the plants on the 1997 IUCN Red List of endangered and threatened species are endemics (Walter and Gillett 1998, xxxiv). For their part, endemic languages small in size and extent (and thus perforce among the most vulnerable to extinction) are so precisely because they have historically adapted to local conditions only. Conversely, as Wurm said, certain species and languages show a great capacity to invade the habitat and social settings of others.[29]

A special kind of invasive species, the organisms involved in epidemic disease, is responsible for the loss of untold numbers of native languages whose speakers were wiped out after European contact. This was (and, in Amazonia, probably still is) an especially important factor in the Americas.[30]

There is also a direct connection between a very small class of favored nonnative species—namely, the domesticated plants and animals used in agriculture—and the death or serious decline of certain languages. As part of a general biocultural deference to the higher-prestige occupations of pastoralism and farming, several hunter-gatherer groups in East Africa have given up their languages in favor of "cattle languages"—those that herders and farmers speak (Dimmendaal 1989, 16–21; Brenzinger, Heine, and Sommer 1991, 31, 39). In Scotland, the infamous Highland Clearances—evictions of peasant farmers—which were tied directly to the desire to increase sheep holdings, helped put Scots Gaelic into the precarious position it is in today (cf. Hamp 1989, 208).

The actions that fragment or destroy wildlife habitat also serve to homogenize cultures and languages: converting wildlands or traditional agroecosystems to intensively managed pasture or cropland (which is in part driven by the burgeoning global export market for agricultural goods); building roads, rail lines, and air strips in remote areas; developing logging concessions, mines, and other industries in indigenous areas, among others. All these encourage a few dominant cultural influences, usually backed by overwhelming political power and money, to spread unchecked.

One of the defining demographic trends of our time is the increasing concentration of the population in cities—a major change in the world's social ecology. The most dramatic shifts to the cities are now taking place in tropical countries, where many endemic species and languages are concentrated. In terms of human psychology, there are profound consequences implicit in switching from a world in which most people grew up in rural areas—close to the land, so to speak—to one in which most grow up in cities (Harmon and Brechin 1994, 107).[31] Might succeeding generations of city-dwellers become so detached from the natural environment that their interest in preserving nature will eventually wane?[32] And could the same apply to the languages and cultures of urban migrants cut off from their ancestral communities? Among those small-language communities in Africa where "migration to the city has become a common pattern," it has been said that "language death is a likely consequence" (Brenzinger, Heine, and Sommer 1991, 32). Some urban migrants may simply reject their former language, as have Nubian speakers in Cairo (Rouchdy 1989, 96), or only the most highly motivated among them will try to keep it up, as has been reported with Irish and Scottish Gaelic (Watson 1989, 45–46).

ASSESSING DIVERSITY

As we saw in Chapter 2, biodiversity is usually evaluated on three fundamental, hierarchically related levels of organization: genes, species, and ecosystems. From an evolutionary perspective, the genetic level is probably the most important because evolutionary change ultimately is reflected in gene frequencies (cf. Futuyma 1989, 573). From a conservation management standpoint, the ecosystem level is arguably the most important because it also includes abiotic (non-living) components of natural communities and thus is the most holistic measurement.[33] In practice, however, biodiversity is usually measured on the species level. Even though counting species merely approximates the range of organic variation in the world, and despite the ambiguities of the species concept itself, "discussion of global biodiversity is typically presented in terms of global numbers of species in different taxonomic groups" (Groombridge 1992, xiii).

Similarly, linguistic diversity can be measured hierarchically, at the structural, the language, and the phylogenetic (or lineage) levels (Table 3.3). Recall that structural linguistic diversity has to do with the amount of disparity exhibited by the structural features and types (e.g., syntax, morphology, phonetics) within a language or group of related languages, while lineage diversity refers to the phylogeny of languages as established by comparative analysis (Nichols 1992, 232, 237).[34] And, though diversity at the language level most properly refers to the range of variation between individual languages across lineages, we saw that the sheer number of distinct languages is almost always used as the proxy for global linguistic diversity, and that it is the most accessible indicator of overall cultural diversity.

Why has the measurement of biological and cultural diversity so consistently fallen back on the rather simplistic expedient of counting species and languages? After all, as we have seen, these two foundation concepts have never been precisely defined. So what lasting claim do they have on us? Wouldn't it be better to wait until off-the-shelf supercomputers allow us to tabulate huge numbers of gene sequences and compare them across organisms?[35] Now *that* would be a real measure of biodiversity. Ditto languages: surely technology is not too far from being able to deliver a hyperfast number-cruncher to every linguist's desktop, allowing him or her to parse out and cross-analyze phonemes and morphemes, the structural bits from which language is built up, using as many sample utterances or writings as he or she wants. Once this kind of high-pow-

Table 3.3
Biological and Linguistic Levels for Assessing Diversity

Biological	Linguistic	Similarities
Genetic	Structural	Scale of resolution: fine
		Potentially most accurate measurement of diversity
		Highly technical concepts; difficult for laypersons to grasp
Species	Language	Scale of resolution: intermediate
		Richness used as proxy of diversity
		Species and languages less well defined than popularly thought
		Intuitive concepts; relatively easy for laypersons to grasp
		Most widely used measures of diversity
Ecosystem	Genetic	Scale of resolution: broad
	(Lineage)	Lack of consensus on how to define ecosystems and lineages; some classifications are thus problematical
		Potentially more accurate measure of diversity than species/ language level

Source: Harmon 1996, 103.

ered, low-cost analytic capability becomes commonplace, it would seem that species and languages could be safely jettisoned as serious scientific concepts.

I have no doubt that technology will soon enough provide us with the requisite wonder boxes. I also have no doubt that species and languages will endure, both in the academic and popular mind. Time and again they have proven resilient in spite of all their ambiguities and deficiencies. The reason is simple: the concepts of species and languages are matched in reality by objectively existing groups of (respectively) natural organisms and cultural behaviors. This makes them, in the well-worn philosophical phrase, "natural kinds." In his essay of that title, W. V. Quine asks, "Why does our innate subjective spacing of qualities accord so well with the functionally relevant groupings in nature as to make our inductions tend to come out right?" The explanation, he finds, is partly neo-Darwinist: "If people's innate spacing is a gene-linked trait, then the spacing that has made for the most successful inductions will have tended to predominate through natural selection. Creatures inveterately wrong in their inductions have a pathetic but praiseworthy tendency to die before reproducing their kind" (Quine 1969, 126). To rephrase and expand:

1. The key to survival is learning to reason from specific examples to general modes.

2. There exist "in nature" (i.e., independently of human thought) real group-
    ings with real significance, rather than mere random agglomerations.
3. Humans have an inborn grasp of how to divide similarities from differ-
    ences: an "innate spacing" (supplemented, I would add, by culturally trans-
    mitted ways of doing it—as recent scholarship in ethnobotanical classifi-
    cation has taught).
4. Because humankind has in fact survived, our inductions must have been
    successful. This means there must be a genuine concordance (which is not
    to say an infallible one-to-one conjunction) between our innate ability to
    recognize similarities and differences and the actual similarities and differ-
    ences that nature presents us with.

Species and languages are examples of such "functionally relevant groupings
in nature" that are concordant with our innate recognitions of spacings (in both
the genetic and cultural senses). That this should be the case is somewhat mys-
tifying. But it is a mystery that can be explained. To do it, we need to next consider some key aspects of how humans construct classifications of the world
and its multiplicity of beings and events.

MONOTHETIC VERSUS POLYTHETIC CLASSIFICATION

Classification systems in Western thought have long lived in the shadow of
Aristotle. His own thinking evidenced two distinct strains in this regard. On
the one hand, he held that every substance (a metaphysical category signifying
that which remains constant in the face of change—a sameness) has at its core
an *essence*, a singular characteristic that gives the substance its "real nature,"
that "makes it what it is" (Sneath and Sokal 1973, 19). In Aristotle's view,
each substance *must* possess the distinguishing essence that it has; it is a matter of necessity. Furthermore, the essence explains the substance's other
characteristics.

On the other hand (as Arthur O. Lovejoy noted), Aristotle also recognized
that "any division of creatures with reference to some one determinate attribute manifestly gave rise to a linear series of classes. And such a series . . . tends
to show a shading-off of the properties of one class into those of the next
rather than a sharp-cut distinction between them. Nature refuses to conform
to our craving for clear lines of demarcation; she loves twilight zones, where
forms abide which, if they are to be classified at all, must be assigned to two

classes at once.[36] And this insensibly minute gradation of differentness is especially evident at precisely those points at which common speech implies the presence of profound and well-defined contrasts" (Lovejoy 1936, 56).

> It will be seen that there was an essential opposition between [these] two aspects of Aristotle's influence upon subsequent thought, and especially upon the logical method not merely of science but of everyday reasoning. There are not many differences in mental habit more significant than that between the habit of thinking in discrete, well-defined class-concepts and that of thinking in terms of continuity, of infinitely delicate shadings-off of everything into something else, of the overlapping of essences, so that the whole notion of species comes to seem an artifice of thought not truly applicable to the fluency, the, so to say, universal overlappingness of the real world. (Lovejoy 1936, 57)

Aristotle therefore was astute enough to see, and candid enough to admit, the shortcomings of relying exclusively on classification by essences. Nonetheless, the concept remained central to his philosophy, and the sharp cut of essentialist classification has been with us ever since, both in the humble reasonings of the common person and in the ruminations of professional philosophers.[37] Using an elaborate symbolic system he called the Logical Alphabet, the 19th-century logician W. Stanley Jevons expounded a "bifurcate classification" in which "each superior class should be divided into two inferior classes, distinguished by the possession and non-possession of a single specified difference." Keep going and one would presumably reach at some point a complete and final stage of knowledge, having discovered the essences of everything. Jevons knew this to be a practical impossibility, of course, but defended the method in principle as being "not only a natural and important one" but "the inevitable and only system which is logically perfect, according to the fundamental laws of thought." Extending the physician's term, he called this method *diagnosis*. Through it, "every class is defined by certain specified qualities or circumstances, the whole of which are present in every object contained in the class, and *not all present* in any object excluded from it." Or, put another way, "whenever a class has been properly formed, a definition must have been laid down, stating the qualities and circumstances possessed by all the objects which are intended to be included in the class, and not possessed *completely* by any other objects" (Jevons 1877, 694, 710, 711).

Such essentialist groups, "formed by rigid and successive logical divisions so that possession of a unique set of features is both sufficient and necessary for membership," have been called *monothetic groups* by modern taxonomists. "They are called monothetic because the defining set of features is unique. That is, all members of any group possess all of the features that are used to define that group" (Sneath and Sokal 1973, 20).

Early systematists in biology—the most famous being Linnaeus—tried to apply monothetic classification to the whole of nature. His binomial nomenclature and hierarchical arrangement of taxa are squarely in the Aristotelian monothetic tradition. Yet even before him, some naturalists had ventured to express their intuitive feelings that this black-or-white approach did not fit the natural world very well. Seventeenth-century botanists such as Pierre Magnol and John Ray felt they had a grasp of what constituted a natural family of plants even though they could not find any single diagnostic character by which to define them (Sneath and Sokal 1973, 20).[38]

As would finally become evident once the mechanisms of natural selection were brought into clear focus, evolution does not produce clean divisions.[39] This has led more recent taxonomists to propose *polythetic* classification as an alternative to the monothetic approach. A polythetic group is any set of entities "in which each entity possesses a large number of attributes of the group, where the attributes might be size, geometric shape, duration of a process, and so forth. Furthermore, each attribute is shared by large numbers of entities, while no single attribute is both sufficient and necessary for group membership"—a radical departure from the monothetic approach (Lumsden and Wilson 1981, 27). Figure 3.2 shows the difference between monothetic and polythetic groups in schematic form.

FOUNDATIONS OF POLYTHETIC CLASSIFICATION

The principles underlying polythetic classification are not new, having been clearly enunciated more than 150 years ago by William Whewell, well known in his day as a philosopher of science. In fact, he was considered by his contemporaries to be something of a polymath: "Science is his forte, and omniscience his foible" pronounced the Reverend Sydney Smith, one of the leading wits of the day. Now Whewell is chiefly remembered for two long treatises, the *History of the Inductive Sciences from the Earliest to the Present Time* (1837) and its double-volume companion, *The Philosophy of the Inductive Sciences, Founded Upon Their History* (1840). Both are filled with decisive utterances and display the kind of assured, unselfconscious sweep of intellect then still thought plausible.

	A	B	C	D	E	F	G	H	I
1	–	–	+	+	+	+	+	+	–
2	+	–	–	+	+	+	–	+	+
3	+	+	–	–	–	–	+	+	+
4	+	+	+	–	–	–	+	–	+
5	+	+	+	+	+	+	+	+	–
6	+	+	+	+	+	+	–	+	+

Figure 3.2. A fully polythetic class clustered around a monothetic core. Attributes are numbered; individuals, lettered. A plus (+) symbol indicates that an individual has the attribute; a minus (–) symbol means the attribute is absent. Attributes 1–6 constitute the determinative set used to ascertain whether the group is a true "class" in Beckner's sense (see discussion in text, below). The subgroup represented by individuals D, E, and F (shaded) is monothetic because each one's complement of attributes is identical. The subgroup represented by individuals A, B, C, and D is fully polythetic because none of their complement of attributes is identical; the same is true of the subgroup represented by individuals F, G, H, and I. The entire group is a true polythetic class because, first, each individual has a large number of attributes in the determinative set (in this example, a minimum of four of the six), and, second, each attribute is possessed by a large number of the individuals (in this example, a minimum of five of the nine). In addition, because no attribute is possessed by *all* of the individuals (attributes 5 and 6 come closest), the class is considered *fully* polythetic. As Beckner (1959) and Sneath and Sokal (1973) point out, in nature very few, if any, taxa are fully polythetic.

Every book ought to have at least one good idea in it. *The Philosophy of the Inductive Sciences* has two: the concept of consilience, in which it is counted a mark of truth when facts from disparate fields converge to reinforce a single conclusion;[40] and that of polythetic classification. Whewell does not use the latter term nor even explicitly state its conditions, but he clearly lays the groundwork for it. Whewell starts his move from square one: the mental act of winnowing sameness from diversity. His opening example is a slant on the old conundrum about the forest and the trees. How, he asks,

> without an exertion of mental activity, can we see one tree, in a forest where there are many? We have, spread before us, a collection of colours and forms, green and brown, dark and light, irregular and straight: this is all that sensation gives or can give. But we associate one brown trunk with one portion of the green mass, excluding the rest, although the neighbouring leaves are both nearer in contiguity and more

similar in appearance than is the stem.[41] We thus have before us one tree; but the unity is given by the mind itself. We see the green and the brown, but we must *make* the tree before we can see *it*. (Whewell 1840, 1:450)

To successfully carve singularity out of the diffuse background of sensation requires that the mental product which is proposed to the consciousness as being singular display *coherence* and *permanence* in meaning (Whewell 1840, 1:451). If I see a collaboration of forms and colors in a forest that I believe to be a tree, I can't be sure that I am correctly apprehending an objective reality unless (1) other people see it too (coherence), and (2) the tree remains a tree, rather than disappearing or metamorphosing into something else (permanence). These criteria are of course easy to satisfy—trivially easy, in fact. And they had better be, since we perform this act of colligation and discernment continuously (at least during our waking hours).[42] Every glance, every odor, every brush of our hand; all the food we taste and sounds we hear; each of our senses, singly or in combination, transmits sensations that our brain instantaneously sorts into samenesses and singularities. These separate entities of perception are then marshaled into coherent scenes within our consciousness through which the individualized action of our life unfolds.

In doing this we have just made use of "the idea of likeness," Whewell says. This process is so fundamental to conscious existence that we can hardly imagine how we would get along without it. "If we had not the power of perceiving in the appearances around us, likeness and unlikeness, we could not consider objects as distributed into kinds at all. The impressions of sense would throng upon us, but being uncompared with each other, they would flow away like the waves of the sea, and each vanish from our contemplation when the sensation ended" (Whewell 1840, 1:452–453). There would be neither coherence nor permanence, and our consciousness (if it could be said to exist at all) would be adrift on the oceans of raw sensation. But we do apprehend likeness amid difference, and this is our anchorage.

So we have separated out our tree and barely had to think about it. But Whewell is not satisfied. He then presses the question of rules: What principle governs the idea of likeness, and, since we use it to make classes of objects, what are the limits of those classes? The answer he gave was a decisive break from essentialist convention:

Perhaps some one might expect in answer to these inquiries a definition or a series of definitions;—might imagine that some description of a tree might be given which might show when the term was applicable and when it was not; and that we might construct a body of rules to which such descriptions must conform. But on consideration it will be clear that the real solution of our difficulty cannot be obtained in such a manner. (Whewell 1840, 1:453–454)

We see an individual tree, we know it to be a tree, and we know that it is a member of a larger class of similar plants called trees, but Whewell is telling us we can't give a black-and-white definition of the class "tree." He is turning his back on monothetic classification, the either-it-is-or-it-isn't school of categorizing. As Jevons did after him, Whewell called this *diagnosis*: the method of using one feature to demarcate a class of individual organisms or objects. Whewell enlarges upon this with the partially opposing term *diataxis*, culled from the jargon of botanical classification, the rules of which, he warned, are at once "curious and instructive" (Whewell 1840, 1:468). Diataxis makes use of latent reference to the nebulous (but, to Whewell's mind, indispensable) concept of "natural affinities" and recognizes that anomalies exist in all useful classifications. Thus, in actuality so-called diagnostic characters do not *constitute*, but rather merely *indicate*, a botanical (or any other) genus. Furthermore, these characteristics "are always subject to be rejected, and to have others substituted for them, when they violate the natural connexion of species which a minute and enlarged study discovers" (Whewell 1840, 1:473).

Just such a deep study reveals few if any clear-cut cases of naturally occurring genera being decided on the basis of pure diagnosis. For example, in the family of the rose-tree, "we are told that the ovules are *very rarely* erect, the stigmata *usually* simple." The "indefiniteness and indecision which we frequently find in the descriptions of such groups," qualities which "must appear so strange and inconsistent" to anyone who imagines they "assume any deeper ground of connexion that an arbitrary choice of the botanist," run contrary to commonsense ideas about how to classify things. But the fact remains that a diagnosis most often "is, though very nearly, yet not exactly commensurate with the natural group: and hence in certain cases this [would-be diagnostic] character is made to yield to the general weight of natural affinities" (Whewell 1840, 1:475).

We are now brought to the cusp of polythetic classification:

These views,—of classes determined by characters which cannot be expressed in words,—of propositions which state, not what happens in all cases, but only usually,—of particulars which are included in a class though they transgress the definition of it, may very probably surprise the reader. They are so contrary to many of the received opinions respecting the definitions of the nature of scientific propositions, that they will probably appear to many persons highly illogical and unphilosophical. . . . But we may here observe, that though in a Natural group of objects a definition can no longer be of any use as a regulative principle, classes are not, therefore, left quite loose, without any certain standard or guide. The class is steadily fixed, though not precisely limited; it is given, though not circumscribed; it is determined, not by a boundary line without, but by a central point within; not by what it strictly excludes, but by what it eminently includes; by an example, not by a precept; in short, instead of Definition we have a *Type*. . . . (Whewell 1840, 1:475–476)

The type is any individual member of a class that is considered to "eminently possess" its characters. This type-specimen forms the nucleus of the species, and so on up the taxonomic scale: "The type-species of every genus, the type-genus of every family, is, then, one which possesses *all* the characters and properties of the genus in a marked and prominent manner," and "though we cannot say of any one genus that it *must* be the type of the family, or of any one species that it *must* be the type of the genus," we can say that "the type must be connected by many affinities with most of the others of its group; it must be near the centre of the crowd, and not one of the stragglers" (Whewell 1840, 1:476–477). In contrast to the strictures governing Aristotle's essences, this is a looser form of necessity.

That way indeterminacy lies, but to Whewell's great credit he did not shun it. In the space of a couple of pages he has fast-forwarded us to the threshold of such ultramodern concepts as fuzzy logic and chaos theory. For "even if there should be some species of which the place is dubious, and which appear to be equally bound to two generic types, it is easily seen that this would not destroy the reality of the generic groups, any more than the scattered trees of the intervening plain prevent our speaking intelligibly of the distinct forests of two separate hills" (Whewell 1840, 1:477). By supplementing diagnosis with diataxis to produce a polythetic classification, we are literally able to distinguish the forest for the trees.

POLYTHETIC CLASSIFICATION: RECENT ELABORATIONS

Thus Whewell staked out the boundaries in the mid-19th century. Darwin grasped the point firmly in the *Origin*: when doing natural history, the value of "an aggregate of characters is very evident" since "a species may depart from its allies in several characters, both of high physiological importance and of almost universal prevalence, and yet leave us in no doubt where it should be ranked." Every classification that has been "founded on any single character, however important that may be, has always failed; for no part of the organisation is universally constant" (Darwin 1859, 417). But polythetic classification did not gain widespread currency until the 1950s (though, again, not under that name) when the man whom many consider the twentieth century's pre-eminent philosopher, Ludwig Wittgenstein, introduced his highly influential "family resemblance" concept. Wittgenstein's discussion of family resemblance is a critical point early on in his *Philosophical Investigations*, published posthumously in 1953. It marks "an ostensible break in the flow of the argument" (Baker and Hacker 1980, 315) in which he has been engaged: the laying of the basis for his philosophy of language by developing the idea of "language-games" and the symbols therein. Then, he brings himself up short:

> Here we come up against the great question that lies behind all these considerations.—For someone might object against me: 'You take the easy way out! You talk about all sorts of language-games, but have nowhere said what the essence of a language-game, and hence of language, is: what is common to all these activities, and what makes them into language or parts of language'. . . . And this is true.—Instead of producing something common to all that we call language, I am saying that these phenomena have no one thing in common which makes us use the same word for all,—but that they are *related* to one another in many different ways. And it is because of this relationship, or these relationships, that we call them all 'language.' (Wittgenstein 1953, para. 65)

As another example, he asks us to consider games. "I mean board-games, card-games, ball-games, Olympic games, and so on. What is common to them all?—Don't say: 'There *must* be something common, or they would not be called "games"'—but *look and see* whether there is anything common to all." Do, he says, and one finds that each comparison brings forth common features, but then some of these drop out in the next round and new commonalities appear "in a complicated network of similarities overlapping and criss-crossing" (Wittgenstein 1953, para. 66).

> I can think of no better expression to characterize these similarities than "family resemblances" [*Familienähnlichkeiten*]; for the various resemblances between mem-

bers of a family: build, features, colour of eyes, gait, temperament, etc. etc. overlap and criss-cross in the same way.—And I shall say: 'games' form a family. (Wittgenstein 1953, para. 67)[43]

The quasitechnical term "family resemblance" (and its derivant, "cluster concept") has been taken up in many fields. I think the kernel of what Wittgenstein was getting at is well expressed by the psychologist of categorization,
Eleanor Rosch, who says his insight is that "we can deal with categories on
the basis of clear cases in the total absence of information about boundaries,"
as when, "in the normal course of life, two neighbors know on whose property they are standing without exact demarcation of the boundaries" (Rosch
1978, 36).[44] Such "clear cases" are what Rosch calls *prototypes*, the best examples
of a class. They are very close to what Whewell meant by "Types," and as such
are sharply to be distinguished from *archetypes*. An archetype has overtones of
immutability, of something given as permanent. It is near in meaning to the
essences of monothetic classification, whereas Rosch's notion of a prototype
is well suited to the polythetic method. More will be said about this shortly.

Wittgenstein's influence on Western thought in the late twentieth century
was unmatched by that of any contemporary biologist.[45] Nonetheless, for truly
rigorous development of the polythetic concept, we must turn back to the field
of biology. The most precise (if not the most graceful) definition of a polythetic
group was that given by Morton Beckner in his 1959 book *The Biological Way
of Thought*.[46] He wrote:

A class is ordinarily defined by reference to a set of properties which are both necessary and sufficient (by stipulation) for membership in the class. It is possible, however, to define a group K in terms of a set G of properties $f_1, f_2, \ldots, f_n$ in a different
manner. Suppose we have an aggregation of individuals (we shall not as yet call them
a class) such that:

1. Each one possesses a large (but unspecified) number of the properties in G
2. Each f in G is possessed by large numbers of these individuals, and
3. No f in G is possessed by every individual in the aggregate.

By the terms of (3), no f is necessary for membership in this aggregate; and nothing
has been said to either warrant or rule out the possibility that some f in G is sufficient for membership in the aggregate. Nevertheless, under some conditions the
members would and should be regarded as a class K constituting the extension of

a concept defined in terms of the properties in G. If n is large, all the members of K will resemble each other, although they will not resemble each other in respect of a given f. If n is very large, it would be possible to arrange the members of K along a line in such a way that each individual resembles his nearest neighbors very closely and his further neighbors less closely. The members near the extremes would resemble each other hardly at all, e.g., they might have none of the f's in common. (Beckner 1959, 22–23)[47]

Beckner's fastidious scientific definition leads to a number of important inferences that are not apparent in a plain-English rendering of the polythetic concept.

First, notice how carefully he refrains from calling his "group K" a *class* from the outset; instead, it is an *aggregate*. An "aggregate" means any old grouping at all, such as those that have been merely thrown together or that are the result of happenstance. A "class" refers to a grouping that has both internal coherence and external significance, in that it "hangs together" in an understandable way while at the same time signifying something true, or at least something useful. The world is full of aggregates of facts that seem pregnant with significance but actually are only coincidences—ask the next conspiracy theorist you happen to meet. Only "under some conditions" (discussed below) does an aggregate become a class. By carefully distinguishing between the two, Beckner sharpens the concept of polythetic groups so that their indeterminacy is meaningful.

Second, Beckner says that there must be a *finite set* (his "set G") of properties that will be determinative of the proposed class. The series f_1, f_2, and so on terminates at some real number, not infinity. This implies that somebody has to decide (or already has decided) which properties "count" toward the determination and which don't. Where to draw the line? The decision is subjective, though not necessarily unscientific for that. Beckner's contemporary, the evolutionary biologist George Gaylord Simpson, repeatedly and unashamedly admitted that taxonomic classification is an art as well as a science, with every taxonomist having to make personal judgments at various places along the line (Simpson 1961, 4, 140, 152, 194, 227).[48] This dictum, it should be noted, applies far beyond biology. For instance, it gets to the heart of how all social groups are formed: the ways by which people collectively identify themselves and subsequently determine whom to accept as a legitimate member of their group.[49]

Third, and most important, Beckner conditions his definition on there being a broad, nearly comprehensive knowledge of the aggregate. In other words, one must know a great deal about a given assemblage of individuals before one can presume to say that they constitute a polythetic class. This may be inferred from the three stipulations he enumerates in the quotation given above. Stipulation (1) amounts to saying that we must know each one of our individuals well, and know how many there are; stipulation (2), that we must have a wide understanding of the properties by which we will assign membership in the putative class; and stipulation (3), that we must put (1) and (2) together (Beckner 1959, 24). As he says, "under some conditions"—namely, those just stated—we are justified in calling the aggregate a class *if* the number of properties being analyzed is large. If we don't know enough about a group to be able to analyze large numbers of its properties, there is no justification for calling it a "class"; rather, it must remain (at least for the time being) a mere "aggregate." The requirement of broad, thorough knowledge—for convenience, let us call it Beckner's Proviso—is the linchpin of the polythetic concept, for by its means the classification being proposed is raised above the level of chance or caprice.

Finally, Beckner observes that one can arrange a very large polythetic class as if it were a chain whose end links hardly resemble one another or may even display no common characteristics at all, yet still are part of a singularity by means of concatenation.[50] Biologists and linguists will instantly recognize real-life examples of this. None other than Darwin himself provides one: "Our classifications are often plainly influenced by chains of affinities. . . . There are crustaceans at the opposite ends of the series, which have hardly a character in common; yet the species at both ends, from being plainly allied to others, and these to others, and so onwards, can be recognized as unequivocally belonging to this, and to no other class of the Articulata" (Darwin 1859, 419).[51] The evolutionary biologist Ernst Mayr has called such "circular overlaps" the "perfect demonstration of speciation" and described in detail a classic example, the circumpolar distribution of the gull *Larus argentatus* (Mayr 1963, 507–512).[52] Numerous other examples from nature have come to light. For their part, linguists recognize their own "chains of affinities" in the phenomenon of dialect chains. These are speech continua in which "geographically adjacent varieties are very similar, whilst those that are separated more widely are much more distinct and may in extreme cases be quite incomprehensible to one another," yet are considered to remain part of a single coherent language (Durrell 1990, 922).

What all this boils down to is that species and languages, though seemingly arbitrary, can still be defined coherently by hewing to Beckner's Proviso and applying polythetic classification methods. Indeed, they are cardinal examples of polythetic groups.

POLYTHETIC GROUPS AS "NATURAL KINDS"

This chapter started with an expression of surprise that biology and linguistics should rest on the same evolutionary tenets. What we've seen since is that the affinities between species and languages are not really surprising at all. As difficult as it has been for biologist and linguists to theorize about them, in the real world they are closely related. They represent actual products of the interplay (stretched out over millennia) between organic and cultural evolution, and are a principal means by which the biocultural presence is registered.

Yet there is a paradox here. Species and languages accord well with our expectations of what is natural. Why, then, do we continually seek after monothetic definitions with seemingly equal conviction? Polythetic entities have a kind of centeredness, it is true: Remember that Whewell referred to them being determined "not by a boundary line without, but by a central point within." But we seem to crave more than this: not just a centeredness, but a *center*.

One of Whewell's most distinguished successors offers the beginnings of an explanation. In the 1920s and 1930s, evolutionary biologists were laying the basis of what became known as the "Modern Synthesis" of Mendelian genetics and the Darwinian theory of natural selection.[53] One of the cornerstones of the Modern Synthesis, Theodosius Dobzhansky's *Genetics and the Origin of Species*, begins with the observation that "for centuries the diversity of living things has been a major interest of mankind":

> Not only are the multitude of the distinct 'kinds' or species of organisms and the variety of their structures seemingly endless, but there is no uniformity within species. . . . From remotest times, attempts have been made to understand the causes and significance of organic diversity. To many minds the problem possesses an irresistible aesthetic appeal, and inasmuch as scientific inquiry is a form of aesthetic endeavor, biology owes its existence in part to this appeal. (Dobzhansky 1941, 3)

Dobzhansky sees the practice of biology as elevated far above the stereotyped image of mannequin men and women in sterile lab coats poring over specimens and peering into microscopes. Biology, as he describes it, is a particular quest: the search for the roots of diversity. What drives certain people to devote themselves to it is the beauty to be found in the looking. There is intense satisfaction to be gained; without it, biology as we know it would not exist.

This unexpected beginning to a landmark book on genetics leads into a model of thinking about diversity, both biological and cultural, in an integrated way. The germ of Dobzhansky's approach is his recognition that nature is not painted in a continuous wash of variety. Rather, "a more intimate acquaintance with the living world discloses a fact almost as striking as the diversity itself," namely, "the discontinuity of the variation among organisms":

> If we assemble as many individuals living at a given time as we can, we notice at once that the observed variation does not form any kind of continuous distribution. Instead, a multitude of separate, discrete, distributions are found. In other words, the living world is not a single array of individuals in which any two variants are connected by unbroken series of intergrades, but an array of more or less distinctly separate arrays, intermediates between which are absent or at least rare.[54] Each array is a cluster of individuals, usually possessing some common characteristics and gravitating to a definite modal point [i.e., statistical center] in their variations. Small clusters are grouped together into larger secondary ones, these into still larger ones, and so on in an hierarchical order. (Dobzhansky 1941, 3–4)

The entire Western scientific taxonomy is built on this last fact, and "evidently the hierarchical nature of the observed discontinuity lends itself admirably to this purpose":

> For the sake of convenience the discrete clusters are designated races, species, genera, families, and so forth. The classification thus arrived at is to some extent an artificial one, because it remains for the investigator to choose, within limits, which cluster is to be designated a genus, family, or order. But the classification is nevertheless a natural one in so far as it reflects the objectively ascertainable discontinuity of variation, and in so far as the dividing lines between species, genera, and other categories are made to correspond to the gaps between the discrete clusters of living forms. Therefore the biological classification is simultaneously a man-made system of pigeonholes devised for the pragmatic purpose of recording observations in a convenient manner and an acknowledgment of the fact of organic discontinuity. (Dobzhansky 1941, 4)

As an example of how this works in practice, Dobzhansky invites us to compare a house cat with a lion. He begins by observing that, as different as individual house cats (and lions) are from each other, no house cat is *so* different that it could be mistaken for a lion, or vice versa. Why? Because "in common as well as in scientific parlance the words 'cat' and 'lion' frequently refer neither to individual animals nor to all the existing individuals of these species, but to certain modal points toward which these species gravitate." While these "modal points are statistical abstractions having no existence apart from the mind of the observer," the species *Felis domestica*[55] and *F. leo* are nevertheless decidedly real, existing independently of "any abstract modal points which we may contrive. No matter how great may be the difficulties in finding the modal 'cats' and 'lions,' the discreteness of species as naturally existing units is not thereby impaired" (Dobzhansky 1941, 4–5).[56] That is to say, whatever troubles there may be in defining what a species is, they are not due to any artificiality of the species themselves.

There is much more going on here than the mere statistical observation that species can be characterized by the bell-shaped "normal curve" most commonly found in the measurement of individual differences (Anastasi 1958, 26–27).[57] Dobzhansky is here expressing a correspondence between our perceptions of what is distinctive—the "abstract modal points"—and what actually *is* distinctive—the "naturally existing units." In so doing he hit upon a fundamental model that explains how diversity is actually structured and how it is perceived, and how certain groups seem to be "natural kinds" in that their perceived modal points anchor the innate mental spacings Quine spoke of (see above section, Assessing Diversity).

For any group entity (a species, a language, a religion, etc.) to be truly discrete, it must encompass some set of characteristics that it does not share exactly with other entities, no matter how closely related they may otherwise be. From an ontological standpoint, this set is definitive. In other words, the complement of characteristics—not taken separately and individually, but as a unique set—is what actually makes the entity distinctive, regardless of whether or not human beings perceive it so. This defining set of characteristics—equivalent to Beckner's finite "set G"—is what coheres around the modal point that people perceive to be there.

Now, this talk of a "definitive" set smacks of the simplistic "either/or" monothetic classification we have already found to be lacking. And, if that were the whole of it, Dobzhansky's model would be just as logical as Jevons' bifurcate classification—and as unrealistic. But Dobzhansky explicitly states that the modal points are mere statistical abstractions. In life, there is no such thing as the archetypal house cat or archetypal lion, though intuitively we think there should be and imagine there could be. Even so, the group entities "house cat" and "lion" are not abstractions: they really exist as biological species because such species are a valid polythetic class. To repeat: For an individual organism to be part of a biological species, it must share with other organisms a large (but unspecified) number of characteristics that have been deemed part of that species' definitive set of characteristics. The delineation of the definitive set must always be subjective, but is nevertheless valid *if* it was made based upon (recalling Whewell's words) "a minute and enlarged study"—which is to say, if it was made by following Beckner's Proviso. Once the definitive set has been thus validly determined, the actual taxonomic judgment—the decision whether to assign the organism to the species—remains subjective, but it is *not* arbitrary. It is a relatively straightforward matter of ascertaining whether a large number of the defining characteristics are present or not. Thus an individual organism can exhibit many disparate, nondefining characteristics and still remain part of a cohesive species. The exact "center" of the Becknerian definitive set—the modal point—exists only in theory and cannot be identified, even while the set itself consists of objectively existing and ascertainable characteristics.

The amazing thing is that the human mind easily leaps over this tangle. Evidently there is something in the way our consciousness functions that allows us to recognize actual discontinuities (Quine's "functionally relevant groupings in nature"). This is axiomatic, of course, and in simple discriminations our fluency is such that the act of recognition is entirely unconscious. But the important point is that we are not invariably stymied by the doubtful cases at the margins; it takes some work, but in most cases we can categorize them. Whether this ability to separate samenesses and differences is something inborn, culturally acquired, or a combination of both is part of a fascinating and perennially controversial debate that we need not join here. Whatever it is, it is natural—or natural-seeming, which is functionally the same thing. It is under this meaning that we can legitimately talk of species and languages being "natural kinds."

It is also apparently natural for humans to construct and accept polythetic classifications of the discontinuities even while, intellectually, leaning toward the use of monothetic methods in the Aristotelian essentialist tradition. As we saw early on in this chapter, it is well known that the criterion of separate gene pools, upon which the classic biological species concept is grounded, is contradicted at various places throughout nature. That being so, one would think that the concept should have been overturned by now. But it instead remains solidly entrenched. The inability of biologists to precisely define a biological species, while at the same time most of them accept it as "the fundamental unit" (Wilson 1992, 37–38), is something of an institutionalized self-reproach within the discipline (see Groombridge 1992, 13; Harmon 1996). It shouldn't be. The endurance of the biological species concept in the face of these seemingly fatal contradictions can be attributed to Dobzhansky's principal insight: that there is a basically accurate correspondence between what we perceive to be separate groups of fundamentally similar organisms and the actual existence in nature of such separate groups.[58] The exceptions to the rule do not disprove it; they merely illustrate the higher order difficulty intrinsic to any analysis of diversity-versus-identity: that of how to separate discrete from continuous variation at the margins, in the gray area away from the modal points.

Dobzhansky's model applies equally well to religions, languages, kinship systems, artistic genres, and the other basic elements of human culture. In all we find versions of the "species problem" cropping up. What, for example, defines the Roman Catholic religion? Is it the doctrines handed down by the Vatican? If so, what do we make of the fact that some of them are regularly contravened by many people who consider themselves (and, importantly, are still considered by the church hierarchy) to be Catholic? When does deviation from doctrine grade into heresy, and from there into total dissolution? These puzzles notwithstanding, no one would deny that Catholicism exists in actuality as a religion. There is evidently a "modal point" of Catholicism, anchoring a definitive set of characteristics around which the meaning "Catholic" coheres. We perceive Catholicism to be a "naturally existing unit" of the world's faiths, and, for all practical purposes, it is.

Working from a premise similar to Dobzhansky's, the scholar of religion John Hick adopted Wittgenstein's family resemblance analogy to explain the plurality of religious traditions alive today. He likens religious diversity to "a complex continuum of resemblances and differences analogous to those found within a family": While no two members of a family are exactly alike, "nevertheless there are characteristics distributed sporadically and in varying degrees which together distinguish this from a different family" (Hick 1989, 4). In other words, faiths are individuated because the variation among them is discontinuous, yet they all are recognizably "religions" in that their practices and beliefs are imbued with a sense of deep, permanent, ultimate importance transcending mundane life. Hick calls this the "starting point" for charting the range of religious phenomena, and it performs the same function for comparative religion as the criterion of reproductive isolation does for taxonomy. The unique elaborations of transcendent importance manifested in different religions become their "modal points," as described in the example of Catholicism above. Using the family resemblance analogy as his platform, Hick builds an elaborate case for the "veridical" nature of religious experience: namely, that faith as it is thought of and experienced verifies an objectively existing "Real"— which, depending on the tradition, takes a personal, theistic form (Islam's Allah, Christianity's Holy Trinity, the God of Israel, etc.) or is conceived of as an impersonal, nontheistic Absolute, such as Nirvana or the Brahman (Hick 1989, 242–245).[59] As Dobzhansky found within the species concept, Hick sees in religious belief a fundamental correspondence between perceived experience and reality. Such correspondence allows us to freely, naturally, even innately apprehend that there are differences among religions even though we may not be able to pinpoint precisely what they are. And, as it does for species, in the end the concept of religion emerges undefined—and unscathed.

Languages too are intuitively considered to be naturally existing units even though linguists have difficulty defining what a language is and how to distinguish among closely related languages and dialects. "Linguistic usage can be described in terms of norms and *ranges of variation* around them," wrote Morris Swadesh. "Each norm represents a central type near which fall the specific instances of a usage. The range of variation is the measure of deflection in one or another direction from the average or most representative form. In some cases the norm is in flux, tending in a given direction, or vacillating between two or more main variants" (Swadesh 1971, 10). Leonard Bloomfield long ago pointed out that "we speak of French and Italian, of Swedish and Norwegian, of Polish and Bohemian as separate languages, because these communities are politically separate and use different standard languages, but the differences of local speech-forms at the border are in all these cases relatively slight and no greater than the differences which we find within each of these speech-communities" (Bloomfield 1933, 54). This is as good a summary of the language-or-dialect problem as one is apt to find. The "oneness" of many languages is geopolitical, not linguistic. For instance, how can there be a single language "Italian" when the Ligurian spoken in the north may well sound like a scramble to a Sicilian (Grimes 1992b, 465–468)? The answer is obvious: at this point in history there happens to exist a bounded entity named "Italy," and it serves many purposes for there to be a single, officially recognized language to match. Away from the centers of power and the linguistic standards they promote, we soon encounter Bloomfield's anomalies. Language or dialect? The dispute—and it is often highly charged—extends to speech forms around the world. Like species, like religions, here again the problem is how to distinguish what is happening at the margins.

Matters of ethnicity are just as perplexing. The neighboring Nuer and Dinka peoples ("Naath" and "Jieng," respectively, in their self-appellations), pastoralists of East Africa, are widely recognized as distinct groups, but the closeness of their languages (divided, though, into greatly differing dialects), the absence of any fixed boundaries between their territories, and their propensity to intermarry all make it very hard to pin down any set of essential distinguishing marks (Bodley 1994, 95–96). Nonetheless, the perception of difference persists among outside observers, and, decisively, the Dinka and Nuer see themselves as being distinctive. Evidently each group recognizes among itself enough shared, identifying characteristics—enough to establish "ethnic modal points"—to keep the two groups separate, no matter how jumbled the boundary between them is.[60]

Finally, there is "culture" itself, a term just as problematical and resistant to easy definition as is "species." There is a conceptual similarity between the two. Dobzhansky observed that variation in the natural world is discontinuous, and built from there. The social anthropologist Fredrik Barth, in his study of the social organization of culture difference, lays the same foundation: "Practically all anthropological reasoning rests on the premise that cultural variation is discontinuous: that there are aggregates of people who essentially share a common culture, and interconnected differences that distinguish each such discrete culture from all others" (Barth 1969, 9). Later, Barth again echoes Dobzhansky when he observes that "the ethnic label subsumes a number of simultaneous characteristics which no doubt cluster statistically, but which are not absolutely interdependent and connected. Thus there will be variations between members, some showing many and some showing few characteristics. Particularly where people change their identity, this creates ambiguity since ethnic membership is at once a question of source of origin as well as of current identity." But this fluidity does not vitiate the basic model:

> What is then left of the boundary maintenance and the categorical dichotomy, when the actual distinctions are blurred in this way? Rather than despair at the failure of typological schematism, one can legitimately note that people *do* employ ethnic

labels. . . . What is surprising is not the existence of some actors that fall between these categories, and of some regions of the world where whole [peoples] do not tend to sort themselves out in this way, but the fact that variations tend to cluster at all. We can then be concerned not to perfect a typology, but to discover the processes that bring about such clustering. (Barth 1969, 29)

So it is not necessary to go so far as to subscribe to Johann Gottfried von Herder's monadic conception of the *Volk*, each endowed with its own "group mind," to believe nonetheless that cultures are objectively existing units, albeit complicated ones. This notion is expressed is some of the most prominent English-language definitions, starting with Edward Tylor, who spoke of culture as a "complex whole," to Edward Sapir's "assemblage" and Clifford Geertz's "pattern of meaning" (cited in Fleischacker 1994, 127–128). Again, it is evident that some definitive set of characteristics is at the center of these complex polythetic unities.

To close this discussion, it should be stressed that one must be careful not to fall into the trap of equating modal points with archetypes, that is, the Aristotelian idea of essences. This is a point that Dobzhansky's fellow evolutionary biologist Simpson insisted upon. From experience in taxonomy he knew that it was all too tempting to take what should be three distinct functions of types and conflate them. The first function is taking a type-specimen to be merely a "vehicle attached to a *name*," that is, as a template for taxonomic nomenclature only, not for taxonomic decision-making. Second, there is the use of type as prototype, as "a standard of comparison, approximation to which warrants *identification* of another specimen as belonging to the same taxon as the type." Last, there is the fallacious function of a type "as the sole or principal basis for the description and *definition* of taxa (primarily of species)"; in other words, erroneously considering a type-specimen to be an essence or archetype (Simpson 1961, 183). This is typology in the old scholastic style, in which it was believed that "the archetype is in some way the reality and that organisms are merely the shadows, the reflections of that transcendental reality"—in short, types as Platonic ideas, as static links in a Great Chain of Being. Simpson vigorously argued that this kind of method is (or should be) thoroughly discredited in modern evolutionary taxonomy (Simpson 1961, 49). He believed that the most pernicious holdover from pre-evolutionary taxonomy is the idea that a single specimen can serve all three functions, whereas "in truly and completely evolutionary taxonomy" the three are totally incompatible (and the last

Table 3.4

Some Analogous Concepts in Biology and Linguistics

Biology	Linguistics	Analogy
Subspecies	Dialect	Variation within basic units of diversity
Species	Language	Basic unit of diversity (cohesion)
Family	Family*	Group of related species/languages
Order	Stock*	Group of related families
Phylum	Phylum*	High-level taxon
Capacity to interbreed or unique combinations of character states	Mutual intelligibility	Simplified definition of species/language: cohesion
Monotypic genus	Language isolate	High importance in diversity
Invasive alien (exotic) species	Politically dominant language	Can displace native species/languages
Lack of reproductive capacity	Moribundity	Loss of capacity to pass genes/language to next generation; the "living dead"
Circular overlaps (Mayr 1963, 507–512)	Dialect chains	Continuous grading of changes producing distinctiveness at end of circle/chain
Cladistics (classification based on evolutionary history)	Language family classification	Use of tree diagrams to illustrate relation among species/languages over long periods of time (but see Dixon 1997)
Refugia (areas where change is or has been retarded)	Residual zones (in the sense used in Nichols 1992)	Areas with accretions of diverse species/languages, including "strandings" (e.g., Hungarian); "residual zones are linguistic refugia where pre-spread stocks regularly survive" (Nichols 1992, 237); cf. Pleistocene refugia

* These terms may be used ambiguously with reference to language (see Ruhlen 1991, 21).
Source: Harmon 1996, 105.

should in any event be relegated to the realm of metaphysics). Taxa in modern taxonomy "are based on samples, not on types" (Simpson 1961, 183; see also pp. 30–31, 49). Final decisions on conspecific status do not depend on a specimen's nearness to a type (in any of the three functional senses of the word), but on whether it falls "within or outside of the ranges of variation *inferred* for the whole taxon," ranges determined by statistical analysis of all specimens previously assigned to the taxon. "The usable sample will of course increase as more specimens are collected and will be different for different classifiers with access to different collections. The proper basis for definition and comparison thus changes and, as a rule, improves as time goes on" (Simpson 1961, 184). This is Beckner's Proviso in action: The more we know about a putative class, the more we can be sure that it *is* a class and not a mere chance aggregation. Furthermore, different (but still valid) samples can be obtained by different persons based on their subjective circumstances.

Here we have a polythetic classification being arrived at through the expansion of the class's boundaries. However, Simpson recognized, in line with Dobzhansky, that "some specimens are of course more nearly average than others as regards particular characters in the sense of being nearer the mean, although this is rarely true of all characters of one organism. The mere fact that a valid average is recognized means that *all* specimens have been taken into account and none especially weighted" (Simpson 1961, 184).

THE REALITY OF SPECIES AND LANGUAGES

This chapter forms the theoretical heart of our exploration of the meaning and value of diversity. The focus has been on species and languages as emblems of biological and cultural diversity. At every turn we have found that comparison between species and languages is not simply metaphorical. It has a basis in real-life shared processes that are reducing their numbers, in actual data on geographical distribution, in a set of analogous concepts (outlined in Table 3.4) that have probative value in explaining diversity, and in their both corresponding to actual discontinuities objectively present in the outer world as well as to perceived distinctions as subjectively registered in the mind.

I want to suggest that this last point endows species and languages with a certain innate validity, an ontological validity, which is unmatched by mere theoretical categories. By bridging perception and reality, and by doing so from such apparently disparate starting points (in accordance, incidentally, with Whewell's concept of consilience), species and languages make an especially rich claim to being carriers of truth. It follows that the realms of diversity they represent can be worked into ethical claims about such critical current issues as globalization, species extinctions, the moral status of nonhuman species, environmental protection, linguistic human rights, and the value of preserving cultural differences.

In Chapter 5 we examine this proposition in some detail. However, before attempting the perilous philosophical jump from description to prescription—going from what *is* to what *ought* to be—we must switch levels again, this time stepping down from global distributions and big abstractions to the plane of individual psychology. Enlisting William James as our guide, in Chapter 4 we look at what people do, mentally, with the diversity experience presents us with. Using a simple model based on James's precepts, we will see how humans construct and use a "sense of sameness" to navigate the whirl of experience which would, without that sense, devolve into chaos. The implications of the model lead to insights about what is at risk, psychologically, if there is a wholesale loss of biocultural diversity. Once those stakes are clear, we can then proceed to a consideration of the moral response to the converging extinction crises. It is through that response that we shall at last encounter the means by which diversity makes us human.

4

WHAT WE DO WITH DIFFERENCE

Comparison is the engine of knowledge. How seldom do we think about the significance of this simple fact, even though, in the last analysis, we spend our mental lives engaged in very little else. As I write this sentence, I . . . stop: to choose the next word from a slate of possibilities, all of them plausible, yet each with a different shade of meaning. The only way to proceed is by comparing the nuances. My gaze drifts from the computer screen up to the pine ceiling. Gradually I find myself conscious of a knot that swirls (a nebula? a whirlpool?) darkly against the honey patina of the pine. The contrast is what engages me at this skimming level of awareness as I churn up possible words. My thoughts shuttle between what I'm looking at and the problem of the unfinished sentence, bringing a vague sense of physical activity going on inside. Finally (and all this may have taken just a few seconds) a leading choice of words comes into the picture, something inside me nods assent, and the sentence proceeds. When I decide to quit writing, it will be to do . . . what? Another cast of choices, with the next move a matter of comparing options.

Now, to rehearse time's passage like this is unspeakably banal unless one can bring the art of a James Joyce to bear upon it. It is obvious that the mechanics of life are an endless round of minor choices with a few momentous ones

119

thrown in. But things aren't quite so straightforward, for the act of comparison has more than a mechanistic role in human thought. If comparing were simply a mathematical or acquisitive felicity, a matter of the brain performing a series of rote operations, there would be no mystery to it at all. Instead, humans process differences in ways that are themselves highly diverse, such that the results of real-life comparisons are often ambiguous and almost always contingent upon other comparisons. Nothing remarkable here, really: as still-evolving social animals, we use a number of tools in our comparison processes (such as instinct, counter-factuality, nonrational reasoning, pure emotion, and the like) that cannot be easily or accurately incorporated into a mechanistic model. For example, we found in the last chapter that humans can be perfectly comfortable with blurry-edged polythetic categories, which is why so few are ready to pitch out species and languages as organizing principles even though we cannot define their evident reality in an either/or way. On the other hand, we also saw that people persist in wanting precisely that sort of monothetically defined category and are willing to invest enormous effort in seeking it out.

We could characterize these tendencies as representing two opposite casts of mind, sometimes all but excluding one another within the confines of a single individual but just as often dissonantly, stubbornly co-dwelling therein. Let us take artists and naturalists as examples of the diversity-affirming mind. Here are creative people who feel awed by the bountiful variety of culture, of nature, or, not infrequently, of both. They derive deep and lasting pleasure from the stimulation that comes with new experiences. They are attracted to variety because it implies a richness of categories and hence of possibilities. To this sort of person, variety is freedom, or at least it enables the drive toward freedom. Differences liberate, they invest the world with expansiveness, they quicken the pulse of life. Nonetheless, no creative person, no matter how bold his sorties into the big, wide world may be, can handle an *overwhelming* amount of diversity. No one likes chaos.

Standing at the other end of the spectrum are those people for whom variety is oppressive, constricting, and threatening. Racists are the obvious example. Yet it is an ironic fact that the persons most disposed to revile human distinctiveness are also those least desirous of eliminating it. The prejudiced mind seeks not to resolve difference, but to maintain and subjugate it. Racists are not really at all interested in seeing this creature of a different skin color become more like themselves. On the contrary, the thought that somehow the gap might be closed (if such an idea occurs at all) fills them with revulsion.

Instead, they wish to calcify the complex bundle of differences that they sense within a neatly ordered personal world-system, a hierarchy in which people of their skin color are at the top. Whatever intellectual failings racists may exhibit, their minds are acutely attuned to, and avidly fond of, difference. They want the difference to be distanced, to be subordinate, but at bottom they want it to remain. Racists, like creative artists or naturalists, have the ability to make fine distinctions and use them as scaffolding to erect a coherent mental structure. Were the result not so odious, the racist's ability would be something to boast of. So, in their very different ways, the artistic, naturalistic, and prejudicial impulses each make use of diversity.

It is clear that people must in some way handle the diversity that the world presents us with. But just *how* do we? Let us begin by affirming that all human beings are in fact able to discern countless differences and that we do so continually all our lives. The bedrock importance of this, "the most fundamental property" of thought, was recognized a century and a half ago by the psychologist Alexander Bain, a man of considerable reputation in his day. The title of Bain's most famous book, *The Senses and the Intellect*, captures the notion that diversity comes to us from nature, arriving through the portals of our senses, while another measure is created within our mind by its conceptual faculty. The senses come first: "To be distinctively affected by two or more successive impressions," he avowed, "may be considered the primary fact of consciousness" (Bain 1904, 335). This he called the "property of discrimination" or "sense of difference." It feeds the very core of the intellect, but then more is added:

> Another fact is *Similarity*, or sense of agreement, which is interwoven with the preceding in all the processes of thought. When we identify any sensation or present mental impression with one that occurred previously, there being an interval between, we exemplify the power of similarity; the sun seen to-day recalls our previous impressions of his appearance. A third fact or property of the Intellect is *Retentiveness*, commonly understood [as] 'memory' and 'recollection.' This power is essential to the operation of the two former powers; we could not discriminate two successive impressions, if the first did not persist mentally to be contrasted with the second; and we could not identify a present feeling with one that had left no trace in our framework. (Bain 1904, 5)[1]

As our senses continuously gather stimuli, these "changes of impression produce at once phenomena of difference and of agreement" in a kind of perpetual dialectic of sensation. "But, although occurring together, the two modes can always be kept separate, and their intellectual consequences run far apart; the one, discrimination, pointing to the individual, the other, agreement, pointing to the general" (Bain 1904, 335).

These are deceptively simple observations. What Bain is in fact describing is the background processing of all that we call "normal thought." As such, it is not surprising that the fact that the world's being is shot through with differences goes entirely unnoticed in everyday mental life. After all, once your house is built, you do not need the blueprints in order to live in it. Bain himself (as he declared everyone did) "took for granted the consciousness of difference as the groundwork of the whole." The "intellectual property of discrimination" is "tacitly assumed in advance as necessary for the exposition of the department of Sensation proper" (Bain 1904, 335–336).[2]

DISTILLING SAMENESS FROM DIFFERENCE: WILLIAM JAMES

Bain thus gave the outlines of human thinking, but it fell to a more illustrious successor in the field of psychology to develop them in a rigorous manner. As a matter of fact, one could fairly say that all of William James's thought, both as psychologist and philosopher, grew out of his conviction that the sense of difference is the primary fact of human consciousness and a leading fact of all existence. With characteristic vigor, James established that real-world diversity and the structure of the mind exist in a state of reciprocity. He will be our guide as we explore "what we do with difference"—the psychological process individuals use to deal with the diversity inherent in the biocultural presence. It is important to understand the process because it is a fundamental and universal aspect of human thought—uniting people as disparate as the artist and the racist, as was suggested above. A close study of James opens up this inner world to us, and provides another basis for appreciating why biocultural diversity is crucial to maintaining a sense of who we are, both as individuals and as part of a greater humankind.

Trained as a physician and physiologist before moving on to establish at Harvard the first psychological laboratory in America, James made his mark in 1890 with the magisterial two-volume *Principles of Psychology*, a work whose influence is still felt today. The leap from there to his later, almost exclusively philosophical works is less than might be supposed, because psychology and philosophy were then much more closely allied than now. As a philosopher,

James is usually cited for his advocacy of pragmatism, a doctrine he borrowed (with attribution) from his colleague C. S. Peirce and then shaped to his own purposes.[3] But James is also one of the most important philosophers to train his sights on questions of identity and diversity, the great ontological fault-lines between the One and the Many, the Same and the Other, the Universal and the Particular[4]—those capitalized Big Dichotomies that undergird philosophy. So impressed was he by these questions that he declared them "the most central of all philosophic problems" (James 1907, 129).[5]

James championed pluralism, the idea that the variety we perceive every day is actually there. This is so much common sense, but, as we found in Chapter 1, Western philosophy since Plato has been dominated by contrary thinking. It has been thought that somehow all the disjunctions appearing to us make sense only when they are rolled into a single, seamless, ineffable, and meta-physically perfect totality. James despised such absolutist and monistic ways of thinking, with their implication that everything has been foreordained. He stood for the messy, the vague, the disjunct, the unfinished, the undeter-mined—for the grit and tumble and freedom of what life is really like.

He wished to validate our daily experience of variety in the world, and demanded a system of thinking that serves that end. "It is surely a merit in a philosophy," he wrote, "to make the very life we lead seem real and earnest. Pluralism, in exorcising the absolute, exorcises the great de-realizer of the only life we are at home in, and thus redeems the nature of reality from essential foreignness. Every end, reason, motive, object of desire, ground of sorrow or joy that we feel is in the world of finite multifariousness, for only in that world does anything really happen, only there do events come to pass" (James 1909, 49–50). Diversity, James believed, is not some projection of human conscious-ness, but rather the very means through which consciousness operates. Of the many traits said to be definitive of humanity, none is more basic than our innate capacity for discerning and classifying difference—and thus resolving portions of it into sameness. James returned to this point again and again, embellish-ing it, recasting it, and burnishing it like a master sculptor.

We are, first of all, born to diversity: "Consciousness, from our natal day, is of a teeming multiplicity of objects and relations, and what we call simple sen-sations are results of discriminative attention, pushed often to a high degree" (James 1890, 1:224). From then until the day we die, the mind is forever choos-ing to attend to one thing or another, but "few of us are aware how incessantly" this faculty is at work because it is the very substance, and therefore seldom also the object, of consciousness.[6] James understood that emphasis is not merely present in all human perception; it is *necessarily* present. A mind saddled with a bland evenness of attention would not long survive in a world of unpredictable hazards and finite resources for subsistence. Focus must take place. In fact, the only way we can deal with great tracts of the variety thrust at us by the world is by ignoring most of the phenomenal terrain (James 1890, 1:284).[7] Luckily, the course of evolution happens to have furnished us with five senses yoked to our consciousness, an interlocking of the senses and the intellect that gives us the consummate ability to divest the world of its primordial confusions. This is fundamental:

> To begin at the bottom, what are our very senses themselves but organs of selection? Out of the infinite chaos of movements, of which physics teaches us that the outer world consists, each sense-organ picks out those which fall within certain limits of velocity. To these it responds, but ignores the rest as completely as if they did not exist. . . . Out of what itself is an undistinguishable, swarming *continuum*, devoid of distinction or emphasis, our senses make for us, by attending to this motion and ignoring that, a world full of contrasts, of sharp accents, of abrupt changes, of picturesque light and shade. (James 1890, 1:284–285)

Endowed with human nature, we are compelled to chisel identity out of objectively existing diversity. We have no choice; that's the way our consciousness is. James called this "sense of sameness" the "very keel and backbone of our thinking," "the most important of all the features of our mental structure" (James 1890, 1:459, 1:460).[8] It is what enables us to confirm that a reality beyond our mind truly exists.

> The judgment that *my* thought has the same object as *his* thought is what makes the psychologist call my thought cognitive of an outer reality. The judgment that my own past thought and my own present thought are of the same object is what makes *me* take the object out of either and project it by a sort of triangulation into an independent position, from which it may *appear* to both. *Sameness* in a multiplicity of objective appearances is thus the basis of our belief in realities outside of thought. (James 1890, 1:272)[9]

Pluralism, unlike absolutism, embraces this fundamental mental act by which we know ourselves to *be* selves.

Yet James also knew that explaining the value of pluralism is difficult. Why? Because "the facts of the world in their sensible diversity are always before us, but our theoretic need is that they should be conceived in a way that reduces their manifoldness to simplicity." The Western tradition of reflective thinking chiefly has been concerned with disentangling the jumble of worldly facts so they can be laid out in a coherent system, such that "philosophers have always aimed at cleaning up the litter with which the world apparently is filled" (James 1909, 45).[10]

> Our pleasure at finding that a chaos of facts is the expression of a single underlying fact is like the relief of the musician at resolving a confused mass of sound into melodic or harmonic order. The simplified result is handled with far less mental effort than the original data; and a philosophic conception of nature is thus in no metaphorical sense a labor-saving contrivance. The passion for parsimony, for economy of means in thought, is the philosophic passion *par excellence*; and any character or aspect of the world's phenomena which gathers up their diversity into monotony will gratify that passion. . . . (James 1897, 65)[11]

"But," James continues, "alongside this passion for simplification there exists a sister passion, which in some minds—though they perhaps form the minority—is its rival. This is the passion for distinguishing; it is the impulse to be *acquainted* with the parts rather than to comprehend the whole."

> Loyalty to clearness and integrity of perception, dislike of blurred outlines, are its characteristics. It loves to recognize particulars in their full completeness, and the more of these it can carry the happier it is. It prefers any amount of incoherence, abruptness, and fragmentariness (so long as the literal details of the separate facts are saved) to an abstract way of conceiving things that, while it simplifies them, dissolves away at the same time their concrete fulness [*sic*]. Clearness and simplicity thus set up their rival claims, and make a real dilemma for the thinker. (James 1897, 66)

A person's philosophic attitude is determined by how he or she balances these competing desires. No system of philosophy, James declared, can hope to be universally accepted if it "grossly violates either need, or entirely subordinates the one to the other."[12] The only possible universal philosophy

> must be a compromise between an abstract monotony and a concrete heterogeneity. . . . A completed theoretic philosophy can thus never be anything more than

a completed classification of the world's ingredients; and its results must always be abstract, since the basis of every classification is the abstract essence embedded in the living fact,—the rest of the living fact being for the time ignored by the classifier. This means that none of our explanations are complete. (James 1897, 66–67)

"We are thus led to the conclusion," he decided, "that the simple classification of things is, on the one hand, the best possible theoretic philosophy, but is, on the other, a most miserable and inadequate substitute for the fulness [sic] of the truth. It is a monstrous abridgment of life, which, like all abridgments is got by the absolute loss and casting out of real matter. This is why so few human beings truly care for philosophy" (James 1897, 69).

Further clouding the picture is an essential paradox, namely, that we cannot have sameness without diversity. James explained it this way:

The perception of likeness is practically very much bound up with that of difference.[13] That is to say, the only differences we note *as* differences, and estimate quantitatively, and arrange along a scale, are those comparatively limited differences which we find between members of a common genus. The force of gravity and the color of this ink are things it never occurred to me to compare until now that I am casting about for examples of the incomparable. Similarly the elastic quality of this india-rubber band, the comfort of last night's sleep, [and] the good that can be done with a legacy, these are things too discrepant to have ever been compared ere now. Their relation to each other is less that of difference than of mere logical negativity. To be found different, things must as a rule have some commensurability, some aspect in common, which suggests the possibility of their being treated in the same way. . . . The same things, then, which arouse the perception of difference usually arouse that of resemblance also. . . . If we start to deal with the things as simply the same or alike, we are liable to be surprised by the difference. If we start to treat them as merely different, we are apt to discover how much they are alike. (James 1890, 1:528–529)

The sense of sameness is absolutely central to our consciousness, but what makes this sense universally human is the *process* of first working through experiential diversity and then distilling categories of sameness from it. Here we have one of the few indisputable universals. Virtually every step in our thinking involves comparison and categorization, and it is by these two means that distillation of sameness from diversity occurs. It is hard to imagine a more diagnostically human trait than the ability, the need, to create and maintain abstract categories of meaning that we mutually recognize as distinctive. This ability is the bottom line of all bottom lines. Without it, language would be impossible. Religion would be impossible. Biological taxonomy would be impossible. Our notion of self would be impossible. Not only does our consciousness of personal identity repose upon a sense of sameness, but some philosophers hold that it is "the only vehicle by which the world hangs together." Certainly "the mind makes continual use of the *notion* of sameness, and if deprived of it, would have a different structure from what it is has . . ." (James 1890, 1:459, 1:460).[14]

DISTILLING SAMENESS FROM DIFFERENCE: A MODEL

Using these principles, we can make a simple model of how humans handle diversity. We begin by affirming with James that diversity truly exists, that it represents or embodies something real, that it is not merely an illusion or a projection of the human mind or that of any other intelligence, and that the universe therefore is not a singular idealistic or monistic entity. If we can agree to this, we can then say that the world is filled with what might be called "raw diversity": separately existing entities, and relations among them, that manifest actual, discrete differences. These entities may be physical things, in which case we can perceive them through our senses, or they may be purely abstract ideas conceived of through our intellect, or, conjunctively, they may be abstract ideas that refer to physical things. Taken all together, these differences make up the field of possible experience. James gets at this idea with his term "perceptual flux," which is simply the ever-changing flow of experience we perceive to be taking place. He repeatedly characterized this flux as a primordial chaos.[15]

Now, on this last point I disagree with James. "Chaos" is a term that denotes utter disorganization. I think, on the other hand, that the primordial condition of the world is more amenable to order than to chaos.[16] Granted, the order that arises is a very complex order. Nonetheless, I see the enduring usefulness of such schema as languages, religions, biological taxonomies, and all the other polythetic entities discussed in the last chapter as prima facie evidence that the world is conducive to order. They speak to the existence of "natural kinds" into which at least some elements of raw diversity individually fall. When taken together, these elements are the full complement of distinguishable, classifiable entities that make up the realms of nature and culture. Raw diversity is, in a sense, the constituent matter of the biocultural presence. Millions of years of evolution have provided humans with an interlocked suite of adaptations which enables us to "process" raw diversity with admirable fluency.

As an example, allow me to imagine myself delivering a talk to a small group of listeners from a lectern in the front of a room. As I take my place and look about, I have an immediate, but wholly subconscious, grasp of my audience's separateness as individuals, of the distinctiveness of the chairs on which they sit, of the plane of the floor, of the dimensional volume of the space we occupy, and so forth. These are all disparate elements, emblematic of raw diversity—quanta of difference, we could call them—yet my experience of the room is nothing like a chaos. Instead, it appears to me as a seamless backdrop, a kind of barely perceived overall experience of "a room."[17] This is how we distill sameness from diversity on the most rudimentary level. Not only do we not need to consciously parse out every single difference that makes up this experiential backdrop, if we were to try to think it out we would become mentally tongue-tied. We would be trying to out-think our own brains, as it were. The human brain has co-evolved with the world's biological and cultural (and especially behavioral) diversity, with the result that it is hard-wired to constantly distill countless sensory stimuli into such seamless experiential backdrops. In so doing—and this is the evolutionary advantage, as I see it—the distillation process infuses our consciousness with a kind of gyroscopic balance, a base feeling of rightness. This gyroscopic rightness can be thought of as the "ground of experience." So, rather than being a chaos, under normal conditions the raw diversity of the world actually forms the unifying, solid foundation of our lived experience.[18]

All this complex sensory processing is happening completely in the background as far as my awareness is concerned. If I were to walk out of the imagined room into bright sunshine and fresh air, the transition would seem perfectly normal because my brain is constantly adjusting to the new stimuli. I would have the sense that a difference had occurred as I moved outside, but I wouldn't have to think about it; it would just "be there." Moreover, under normal circumstances I would perceive the change as a continuous flow and not as a ratcheting series of discrete moments in space and time. That, of course, is the notion captured in James's most famous conception, the stream of thought.[19] When he introduced the image in the *Principles*, James was reacting against the then-prevalent associationist school of psychology, which taught that complex ideas were just built up from simpler ones and that certain ideas could call up or inhibit others, thereby governing mental life in an atomistic, rigid manner (Perry 1936, 2:58, 2:76–80). Destitute as imagery, impoverished as explanation, associationism held no appeal for James. Rather, "what must be admitted is that the definite images of [this] traditional psychology form but the very smallest parts of our minds as they actually live."

> The traditional psychology talks like one who should say a river consists of nothing but pailsful, spoonsful, quartpotsful, barrelsful, and other moulded forms of water. Even were the pails and the pots all actually standing in the stream, still between them the free water would continue to flow. It is just this free water of consciousness that psychologists resolutely overlook. Every definite image in the mind is steeped and dyed in the free water that flows round it. With it goes the sense of its relations, near and remote, the dying echo of whence it came to us, the dawning sense of whither it is to lead. The significance, the value, of the image is all in this halo or penumbra that surrounds and escorts it,—or rather that is fused into one with it and has become bone of its bone and flesh of its flesh; leaving it, it is true, an image of the same *thing* it was before, but making it an image of that thing newly taken and freshly understood. (James 1890, 1:255)

To return to our imaginary example, my grasp of the overall ambiance of the room is the definite image; and, as James correctly understood, it forms only the smallest part of my mental landscape at that particular moment. Without its entourage—concurrent perceptions of my audience as individuals, of the volume of the room, of the plane of the floor, and so on, not to mention the whole complex of thoughts swirling around whatever remarks I am trying to convey to my listeners—without all this, my image of the room could never take shape. And all the while I barely perceive this coming-together as "something happening." I say "barely perceive" even though at this stage I am really operating on a preperceptual level because (and it sounds silly to say it) I have no sense that my senses are involved. When we are in this "instant field of the present" we operate in what James called a "world of pure experience" (James 1912, 23). Unconstrained by the intellect, "experience in its immediacy seems perfectly fluent. The active sense of living which we all enjoy, before reflection shatters our instinctive world for us, is self-luminous and suggests no paradoxes. . . . When the reflective intellect gets at work, however, it discovers incomprehensibilities in the flowing process. Distinguishing its elements and parts, it gives them separate names, and what it thus disjoins it cannot easily put together" (James 1909, 347).

And it is to the reflective intellect that we now turn for the second step in our model. For we humans are not just bundles of nerves soaking up stimuli; we are also equipped with volition and the capacity to reason in the abstract. In other words, we act on perceptions according to our own purposes and interests, which are themselves transformed into conceptions arranged into a network of categories within our minds.[20] "The phenomena of selective attention and of deliberative will are of course patent examples of this choosing activity. But few of us are aware how incessantly it is at work in operations not ordinarily called by these names. Accentuation and Emphasis are present in every perception we have. We find it quite impossible to disperse our attention impartially over a number of impressions" (James 1890, 1:284, 1:461). So, the second step in distilling sameness from diversity is purposeful: Through accentuation and emphasis we move (often rapidly, sometimes instantly) from a preperceptual comprehension of difference to the first, basic, intellectualized conceptions of difference.[21]

So once volition and conception elbow their way into the picture, the native "flowingness" of experience begins to be disrupted. Let us return to the example of my experience of giving a talk to an audience. If my attention becomes focused on a specific part of the room, say, on someone's face, suddenly my thoughts catch. A host of half-realized thoughts are likely to run through the back of my mind: Who is this person? Have I met him before? What would I say to him if we were introduced? And so on. He has become individualized for me, and my experience of the room is, all of a sudden, no longer unconscious and unified and fluent, but conscious and "broken up" and halting.

The third step in the model represents a more sophisticated intellectual plane by far. Here we go from basic conceptions of difference to those that compare the magnitude of and quality of difference. It is at this stage that the mind builds complex categories. The reason we can do this is because our minds are able to hold onto samenesses—which we call "meanings"—and recall those meanings through memory. Not only are meanings permanent mental possessions, "but the results of their comparison are permanent too. The objects and their differences together form an immutable system. The same objects, compared in the same way, always give the same results. . . ." Moreover, "we can think of the various resemblances and differences which we find and compare them with each other, making differences and resemblances of a higher order. The mind thus becomes aware of sets of similar differences, and forms series of terms with the same kind and amount of difference between them . . ." (James 1890, 2:644–645). As we pass from childhood to maturity, and gain in experience and education, we use this comparative facility to create a network of finely graded classifications.[22]

It is a remarkable fact that some of the most sophisticated and effective classifications are those that cannot be defined precisely. We saw in Chapter 3 that species and languages are outstanding examples. Everyone makes wide use of these notions, but when biologists and linguists try to define exactly what they are, the result is confusion and disagreement (Harmon 1996). Despite that, both experts and laypersons recognize species and languages as not only useful categories, but also entities with a real existence. They seem to us to be natural kinds. And thus, precisely because they bring with them a feeling of naturalness, at this supreme level of classificatory sophistication we regain some of the fluency that was lost after we moved beyond the first, preperceptual level of the model.

It is no exaggeration to say that virtually our entire mental life is based on iterations, in whole or in part, of this three-stage process of comparison-making.[23] By sifting and re-sifting through all this diversity, we wind up by creating an individual worldview for ourselves. "The mind," wrote James, "works on the data it receives very much as a sculptor works on his block of stone. In a sense the statue stood there from eternity. But there were a thousand different ones beside it, and the sculptor alone is to thank for having extricated this one from the rest."

> Just so the world of each of us, however different our several views of it may be, all lay embedded in the primordial chaos of sensations, which gave the mere *matter* to the thought of all of us indifferently. We may, if we like, by our reasonings unwind things back to that black and jointless continuity of space and moving clouds of swarming atoms which science calls the only real world. But all the while the world *we* feel and live in will be that which our ancestors and we, by slowly cumulative strokes of choice, have extricated out of this, like sculptors, by simply rejecting certain portions of the given stuff. Other sculptors, other statues from the same stone! Other minds, other worlds from the same monotonous and inexpressive chaos! My world is but one in a million alike embedded, alike real to those who may abstract them. (James 1890, 1:288–289)[24]

DIVERSITY AND DETERMINISM

This, then, is a simple and very general model of how humans handle the diversity that our experiences bring to us.[25] If taken as a purely cognitive model, it would probably be of little value, having long since been surpassed in sophistication by the work of James's successors in the fields of psychology. But James never would have claimed such a narrow province for the use of his ideas. The years of effort and drudgery that went into the writing of *The Principles of Psychology* were more than enough to sate his interest in experimental and theoretical psychology for the rest of his life.[26] What James was instead interested in, almost as if constitutionally, was the borderland where psychology and philosophy come together. The interest is marked in all his post-*Principles* philosophical work, but it reached its deepest expression in his classic analytical-study-cum-testament, *The Varieties of Religious Experience*, which was published in 1902. Anyone who reads that renowned book will come away with the understanding that James could never be satisfied with a blind reliance on a single way of knowing, whether it be the way of science, theology, secular philosophy, or anything else.[27] His intellectual restlessness he seemed to share more with his father than with his famous brother, but as it was for the younger Henry James, one of William's great themes was exploring the behavior of choosing and the momentousness of choice. William expressed their importance in terms of free will, of undictated volition as the basis for our decisions being truly consequential.

He came to that philosophical place only after having traveled a very hard road. In his early manhood he went through a prolonged and severe depression which, once he finally emerged from it, proved to be a course-setting event. James's malaise was caused by years of brooding over his lack of a guiding philosophy at once deep and flexible enough to stand up to the implacable fact of the existence of evil. Gratuitous evil, evil so heinous as to be beyond all comprehension, evil in unspeakable amounts—these were live issues for James. They importuned him, ate at him, compelled him to seek alternatives far beyond conventional theological moralizing. It *was* compulsion; as his biographer, Ralph Barton Perry, described it, "He was looking for a solution that should be not merely tenable as judged by scientific standards, but at the same time propitious enough to live by. Philosophy was never, for James, a detached and dispassionate inquiry into truth; still less was it a form of amusement. It was a quest, the outcome of which was hopefully and fearfully apprehended by a soul on trial and awaiting its sentence" (Perry 1936, 1:323). What finally brought James out of the depths was his embrace of freedom (and all its attendant possibilities of error and disorder) as the motive force in human affairs. Inspired by the writings of the philosopher Charles Renouvier, James made his own leap of faith: "My first act of free will," he vowed to his diary on April 30, 1870, "shall be to believe in free will" (Perry 1936, 1:323). If there was one indispensable facet of James's many-faceted personality, it was his hatred of determinism.[28] Keeping this in mind helps clarify his entire approach to psychology and philosophy.

Under James's terms of reference, the first stage of the model sketched above is an explanation of free will as it operates on the most basic level within the human mind. At this level, there is indeed little to distinguish it from what the evolutionary biologist George Gaylord Simpson saw as a universal property belonging to all species:

> The necessity for aggregating things (or what is operationally equivalent, the sensations received from them) into classes is a completely general characteristic of living things. One should hesitate, although some do not, to apply such words as 'consciousness' or 'perception' to an amoeba, for instance, but it is perfectly obvious from the reactions of an amoeba that something in its organization performs acts of generalization.[29] It does not react to each bit of food, say, as a unique object, but in some way, in some sense of the word, *classifies* innumerable different objects all within the class of foodstuffs. Such generalization, such classification in that sense, is an absolute, minimal requirement of adaptation, which in turn is an absolute and minimal requirement of being or staying alive. (Simpson 1961, 3)[30]

Thus is human behavior, at this most fundamental level, linked with that of every other animal on the planet. But of course complex volition—or, if you prefer, the will— is what takes *Homo sapiens* far above and beyond the deterministic ground floor occupied by amoebas and similar creatures.[31] People cannot will new material things into existence out of nothingness, but we can will new ideas and ways of living into being, and these often do set off chains of events that eventually lead to profound change in the material world. The uniquely human capacity to conceptualize higher order abstractions and then willfully enact them (the second and third stages of the model) is the ultimate source of the power of our species over the rest of the planet. It is a power not to be underestimated, and so sometimes, in the heat of argument, James unguardedly spoke as if it were sovereign: "Talk of logic and necessity and categories and the absolute and the contents of the whole philosophical machine-shop as you will, the only *real* reason I can think of why anything should ever come is that *some one wishes it to be here*" (James 1907, 288–289).[32]

Such statements testify to the fact that James wanted nothing more than to drive absolutism from the field absolutely, not just in secular philosophy:

> Whoever claims *absolute* teleological unity, saying that there is one purpose that every detail of the universe subserves, dogmatizes at his own risk. Theologicians who dogmatize thus find it more and more impossible, as our acquaintance with the warring interests of the world's parts grows more concrete, to imagine what the one climacteric purpose may possibly be like. We see indeed that certain evils minister to ulterior goods, that the bitter makes the cocktail better, and that a bit of danger or hardship puts us agreeably to our trumps. We can vaguely generalize this into the doctrine that all the evil in the universe is but instrumental to its greater perfection. But the scale of the evil actually in sight defies all human tolerance; and transcendental idealism . . . brings us no farther than the book of Job did—God's ways are not our ways, so let us put our hands upon our mouth. A God who can relish such superfluities of horror is no God for human beings to appeal to. (James 1907, 142–143)[33]

Unlike absolutists, James was satisfied with a "concatenated knowing," one which goes "from next to next." It makes for "a coherent type of universe in which the widest knower that exists"—God, as many would have it—"may yet remain ignorant of much that is known to others." This is "altogether different from the 'consolidated' knowing supposed to be exercised by the absolute mind" (James 1911b, 129–130).[34] The "widest knower" in this manifestly polythetic universe can be thought of as the ultimate embodiment of Beckner's Proviso. Yet even this being, however we might characterize it, does not know everything. Omniscience is not possible.

James, I hasten to say, defended the right to believe, including the right to believe conventional beliefs, and his reputation has come under fire from certain philosophical gun-emplacements because of it. Yet he never advocated willy-nilly belief. He instead demanded that theology comport itself in accordance with the palpable realities of the universe. It is just that the palpable realities are not the only ones: beyond Earth, James believed along with the absolutists, is some kind of divinity, a divine substance. "But whereas absolutism thinks that the said substance becomes fully divine only in the form of totality, and is not its real self in any form but the *all*-form, the pluralistic view which I prefer to adopt is willing to believe that there may ultimately never be an all-form at all, that the substance of reality may never get totally collected, that some of it may remain outside of the largest combination of it ever made, and that a distributive form of reality, the *each*-form, is logically as acceptable and empirically as probable as the all-form commonly acquiesced in as so obviously the self-evident thing" (James 1909, 34). The "strung-along unfinished world" (James 1909, 128) thus does not forbid humanity from connecting with divinity; it simply concedes that divinity may never come in the form of a monothetic whole. By refusing to accept that conclusion, the defenders of absolutism fall prey to the "architectonic instinct," as invoked by George Santayana in a contemporaneous review of *The Principles of Psychology*:

> Those who deal with the abstract and the general, who think impersonally and along the lines of a universal system, are almost sure to ignore their own ignorance. They acquire what has been called the architectonic instinct; their conceptions of things are bound to be symmetrical and balanced, and fit into one another with perfect precision. . . . Their cold breath congeals the surface of truth into some system; and on that thin ice they glide merrily over all the chasms in their knowledge. (cited in Perry 1936, 2:110)[35]

A POLYTHETIC UNIVERSE

Clearly, such reactions against thoroughgoing determinism and system-mongering are closely tied to James's rejection of absolutism and monism in metaphysics. He complained that "the commonest vice of the human mind is its disposition to see everything as yes or no, as black or white, its incapacity for discrimination, of intermediate shades" (James 1909, 77–78). Those who hold absolutist doctrines are guilty in the extreme of this overly exclusionary, overly monothetic approach. They "talk as if the minutest dose of disconnectedness of one part with another, the smallest modicum of independence, the faintest tremor of ambiguity about the future, for example, would ruin everything, and turn this goodly universe into a sort of insane sand-heap or nulliverse, no universe at all"; as if "the unity of things and their variety belonged to different orders of truth and vitality altogether" (James 1897, 154–155; James 1912, 44).[36]

James taught the opposite. The universe is pluralistic, and it holds together by the same means ring species[37] and dialect chains do: through concatenation, the overlapping of partially similar entities, the "through-and-through union of adjacent minima of experience, of the confluence of every passing moment of concretely felt experience with its immediate next neighbors." Such contiguity produces "a coherent world, and not an incarnate incoherence, as is charged by so many absolutists. Our 'multiverse' is still a 'universe'; for every part, tho [*sic*] it may not be in actual or immediate connexion, is nevertheless in some possible or mediated connexion, with every other part however remote, through the fact that each part hangs together with its very next neighbors in inextricable interfusion" (James 1909, 325–326).

In concatenated knowing, thought proceeds by means of a "magical, imponderable streaming" in which "trains of imagery and consideration follow each other" in a "restless flight of one idea before the next. . . ." Leaping and darting, dodging and twisting, our thoughts pass "between things wide as the poles asunder"; yet these transitions, "which at first sight startle us by their abruptness . . . when scrutinized closely, often reveal intermediating links of perfect naturalness and propriety . . ." (James 1890, 1:550).[38]

With its links and intermediaries, its immense depth and reach, Jamesian concatenation seems to recall the central features of the Great Chain of Being. But the differences between them are basic. In the Chain of Being, every position is predetermined, static, perfectly defined, and in a linear hierarchy ordered according to an entity's supposed value as measured by its nearness to God. James's multiverse is dynamic,[39] vague, nonlinear (more like Darwin's tangled stream bank than a forged chain), and not conditioned by fixed value distinctions. The unsystematic subjectivity of individual personalities fills in the gaps between purposive activity. James speaks of there being "nuclei of shared reality" round which "floats the vast cloud of experiences that are wholly subjective . . . the mere day-dreams and joys and sufferings and wishes of the individual minds. These exist *with* one another, indeed, and with all the objective nuclei, but out of them it is probable that to all eternity no interrelated system of any kind will ever be made" (James 1912, 65–66). All but limitless, this disorganized mind work has no apparent purpose:

Most people probably fall several times a day into a fit of something like this: The eyes are fixed on vacancy, the sounds of the world melt into confused unity, the attention is dispersed so that the whole body is felt, as it were, at once, and the foreground of consciousness is filled, if by anything, by a sort of solemn sense of surrender to the empty passing of time. In the dim background of our mind we know meanwhile what we ought to be doing: getting up, dressing ourselves, answering the person who has spoken to us, trying to make the next step in our reasoning. But somehow we cannot *start*; the *pensée de derrière la tête* fails to pierce the shell of lethargy that wraps our state about. Every moment we expect the spell to break, for we know no reason why it should continue. But it does continue, pulse after pulse, and we float with it, until—we know not what—enables us to gather ourselves together, we wink our eyes, we shake our heads, the background-ideas become effective, and the wheels of life go round again. (James 1890, 1:403–404)

James described the mind's operations in terms very much like the polythetic conceptions we discussed in Chapter 3. His "nuclei of shared reality," reminiscent of Dobzhansky's modal points, are items about which there is very broad agreement, and which carry adaptive value in an evolutionary sense. They provide the anchor points for social intercourse. Similarly, within ourselves as individuals the course of thought is marked by a succession of "mental fields," each with its own "center of interest" to which we pay close attention, "around which the objects of which we are less and less attentively conscious fade to a margin so faint that its limits are unassignable."

Some fields are narrow fields and some are wide fields. Usually when we have a wide field we rejoice, for we then see masses of truth together, and often get glimpses of relations of truth together, and often get glimpses of relations which we divine rather than see, for they shoot beyond the field into still remoter regions of objectivity, regions which we seem rather to be about to perceive than to perceive actually. At other times, of drowsiness, illness, or fatigue, our fields may narrow almost to a point, and we find ourselves correspondingly oppressed and contracted. . . . The important fact which this 'field' formula commemorates is the indetermination of the margin. Inattentively realized as is the matter which the margin contains, it is nevertheless there, and helps both to guide our behavior and to determine the next movement of our attention. . . . Our whole past store of memories floats beyond this margin, ready at a touch to come in; and the entire mass of residual powers, impulses, and knowledges that constitute our empirical self stretches continuously beyond it. So vaguely drawn are the outlines between what is actual and what is only potential at any moment of our conscious life, that it is always hard to say of certain mental elements whether we are conscious of them or not. (James 1902, 231–232)

James was a master at distinguishing and describing mental states—actual mental states, the kind which lead to actual behavior. His empirical convincingness is what puts the brand of truth on his conception of a concatenated universe. Rather than try to spin out an *a priori* explanation of things, as did the creators and defenders of the Great Chain of Being, James tells us what he sees. The price of this method is indeterminacy, vagueness, and open-endedness, and all the potential for anxiety that comes with them. For his part, James was gladly willing to pay.[40] For the reward is escape, escape from the kind of lock-step existence implied by the Chain of Being. And this brings us to the most important distinction of all, which is that activity in James's universe is not pre-determined. Just the opposite: It depends on there being real freedom and con-tingency, in earthly actions counting for something. If existence is truly like this—if it is really concatenated, pluralistic, and largely polythetic—then, "so far from defeating its rationality, as the absolutists so unanimously pretend, you leave it in possession of the maximum amount of rationality practically attain-able by our minds. Your relations with it, intellectual, emotional, and active, remain fluent and congruous with your own nature's chief demands" (James 1909, 319).

Fluency and congruity come from engaging both the monothetic and poly-thetic aspects of existence. Concepts are monothetic, exclusively so, and can-not suffice on their own. "When we conceptualize, we cut out and fix, and exclude everything but what we have fixed. A concept means a *that-and-no-other*. Conceptually, time excludes space;[41] motion and rest exclude each other; approach excludes contact; presence excludes absence; unity excludes plu-rality; independence excludes relativity; 'mine' excludes 'yours'; connexion excludes that connexion—and on indefinitely; whereas in the real concrete sen-sible flux of life experiences compenetrate [interpenetrate] each other so that it is not easy to know just what is excluded and what not" (James 1909, 253–254). The flux of life, to use our term, is polythetic: "Our fields of expe-rience have no more definite boundaries than have our fields of view. Both are fringed forever by a *more* that continuously develops, and that continuously supersedes them as life proceeds" (James 1912, 71). The actual "pulses of expe-rience appear pent in by no such definite limits as our conceptual substitutes for them are confined by. They run into one another continuously and seem to interpenetrate" (James 1909, 282). "Here, then, inside the minimal pulses of experience, is realized the very inner complexity which the transcendentalists say only the absolute can genuinely possess"; and, as a matter of fact, "the whole process of life is due to life's violation of our logical axioms" (James 1909, 284, 257).

So concepts in their monothetic splendor are the vehicle by which humans surmount mere deterministic choice-making, yet if we rely too much on conceptualizing we drain life of its vitality. Indeed, one could go so far as to say that life itself, taken in its broadest outlines, is polythetic because it subsumes all monothetic entities beneath an overarching, coherent diversity. James therefore has an age-old answer to the age-old question of whether the essence of things is one or many: it is both. The universal and particular parts of experience "are literally immersed in each other, and both are indispensable"; in the end, "neither the oneness nor the manyness seems the more essential attribute" for "they are coordinate features of the natural world" (James 1911b, 107, 126–127).[42]

LOSS OF DIVERSITY: COGNITIVE AND CULTURAL CONSEQUENCES

To those who are unsympathetic with his aims and methods, it is easy enough to paint James as a mere historical figure, someone who played a large part in the development of a nonderivative intellectual life in America, a pioneer, colorful and witty and still readable, but necessarily outdated in terms of substance and of no real pertinence for today. Nothing could be more mistaken. Far from being part of the wooden furniture of academic history, James's theory of how the mind works within a pluralistic universe is keenly relevant as society struggles to come to terms with globalization and the post-Cold War world order. Better than anyone else, James recognized the ambiguities inherent in what humans do with sameness and difference. He thought speculatively and with unmatched insight about the psychology of that process, paving the way for subsequent studies of categorization that have enlarged our understanding of how worldviews are ordered among individuals across various cultures. Fundamentals of thought: They are what James plumbed, and surely they are what will ultimately determine how we react, singly and collectively, to the current upheavals in the world's economy and environment.

James laid the groundwork, and because of him and those who followed we can consider a critical question: What are the possible long-term psychological consequences of a wholesale loss of biological and cultural diversity?

First, we see that as elemental differences are erased (e.g., through species extinction and language destruction), the field of possible experience becomes more threadbare, more constricted. As time goes on, and successive generations are exposed to less "diversity bandwidth," our innate ability to distill sameness from diversity will begin to atrophy.[43] Left unchecked, globalization and extinctions will give us sameness in abundance, but it will be sameness handed to us on a silver platter—or perhaps we should say in a paper wrapper, since what we we're talking about here is the principle behind McDonald's. And to be sure, a steady diet of Big Macs is not conducive to developing a discriminating palate. If diversity and our ability to discern it are radically reduced, so too will be our readiness to embrace or even conceive of alternative viewpoints to challenge any particular proposed course of action. That would make the human race even more vulnerable to charismatic, totalitarian manipulation.

There are other risks. Bryan G. Norton makes an observation about species that applies equally well to languages and cultures:

> In a world increasingly dependent upon monocultures, there will be not only a decrease in the area covered by complex, naturally evolved, and efficient systems . . . but also *a decrease in the possibility of such systems emerging in the future.*[44] If present trends continue, monocultures will cover much of the land area of the globe, and the local and worldwide extinctions caused by such practices will undermine the total diversity of local areas and of the world as a whole. The resulting impoverished world will lack the building blocks for the reconstruction of such systems. The use of humanly enforced monocultures produces a spiraling trend: *as more and more space is so occupied, there will be less and less possibility of reversing the trend* by reinstituting unmanaged areas and allowing them to evolve into complex, efficient systems. (Norton 1987, 91, emphases added; see also pp. 206–207)

Whether in nature or society, introduced monocultures eliminate diversity—irreversibly if the process is carried far enough so that the result is extinction. The potential psychological effect on humankind of such cumulative simplification is enormous. "Some experiences," observed James, "simply abolish their predecessors without continuing them in any way" (James 1912, 62). Extinction is the quintessential terminal experience. Not only is a life-form or cultural entity lost, but the loss itself abolishes the opportunity for future generations to directly experience what is gone. No one can reasonably argue that looking at a stuffed passenger pigeon in a museum or reading a dry description of the extinct language Tasmanian is equivalent to seeing the bird alive or hearing the language spoken. Nor will so-called virtual reality, delivered to one's computer, ever match real reality, no matter how sophisticated it becomes in replicating the interplay of the five senses and simulating spatial movement. The reason is simple: real experiences are unscripted and always carry with them the possibility of novel twists and turns that cannot be anticipated. By definition, virtual reality must be programmed, and all programming has finite bounds that are much more circumscribed than any found in unmediated nature and culture.

Extinction also breaks the evolutionary continuity we are heir to, producing a deep-seated disruption that alienates us from the biocultural presence in a downward spiral: the more simplification, the more extinction, the more estrangement from the diversity that remains. For children growing up in a depauperate landscape—that is, one devoid of its former natural and cultural variety—the experience of life has no felt connection with that of their ancestors. What is perfectly normal to them would have been abnormal to the hundreds of generations that came before. It is extremely difficult for them to feel the loss of something they've never known firsthand. On the other hand, the continuation of biocultural diversity allows coming generations to enlarge upon their experiences of their ancestors. Instead of terminating, intergenerational experience keeps evolving. To continue diversity is truly to enlarge the meaning of living experience by providing increasing scope for verifications that lead to—of all things—a more unified and consolidated understanding.[45] Under conditions of continuity, as James realized, "the unity of the world is on the whole undergoing increase" even though the temper of that world remains pluralistic. "The universe continually grows in quantity by new experiences that graft themselves onto the older mass; but these very new experiences often help the mass to a more consolidated form" (James 1912, 89–90). But if the continuity between old and new experiences is severed, this building-up of unity ceases.

If we extend Norton's areal "space"[46] and "management" metaphors to culture, we might say that as more and more of the collective cultural–psychological space of the world's peoples is taken up with narrower and narrower mental constructs—"monocultures of the mind," as Vandana Shiva calls them (1993)—there will be less and less capacity to generate alternative constructs. For these ever-spreading monocultures of the mind are "managed" areas of the mental landscape, created as the result of incessant advertising, political propaganda, and similar forms of manipulative (and, now, increasingly pecuniary) thought production.[47] Unlike previous periods in history, it is now difficult to avoid being exposed to them. By taking up an increasing amount of our time—that is, more of the spatiotemporal continuum that characterizes thinking and consciousness—managed, manipulated thought literally crowds out unmediated forms of thinking. Yet it was just this sort of undirected (or, at most, only vaguely directed) thinking that produced most of the cultural content that the world has ever had, and hence the "complex, efficient systems" of the biocultural presence. The crux of this argument is that once a certain critical portion of the collective cultural–psychological landscape is colonized, it becomes extremely difficult to "reclaim" it for the purpose of unmediated thinking— just as hard as it is to, say, restore Iowa corn fields into native prairies.

Perhaps even harder. People know (or are readily able to learn) what a genuine prairie looks like, and can easily tell it from a corn field. But, as the political scientist Benjamin Barber astutely points out, today's global marketing strategy "depends on a systematic rejection of any genuine consumer autonomy or any costly program variety—deftly coupled, however, with the appearance of infinite variety."

> Selling depends on fixed tastes (tastes fixed by sellers) and focused desires (desires focused by merchandisers). Cola companies . . . can no more afford to encourage the drinking of tea in Indonesia than Fox Television can encourage people to spend evenings at the library reading books they borrow rather than buy; and Paramount, even though it owns Simon & Schuster, cannot really afford to have people read books at all unless they are reading novelizations of Paramount movies. By the same logic, for all its plastic cathedrals, Disneyland cannot afford to encourage teenagers to spend weekends in a synagogue or church or mosque praying for the strength to lead a less materialistic, theme-park-avoiding, film-free life. Variety means at best someone else's product or someone else's profit, but cannot be permitted to become no product at all and thus no profit for anyone. (Barber 1995, 116)

The burgeoning global consumer culture, then, depends upon a sham diversity that presents trivial choices (which brand of paper towel is more absorbent?) as if they were important. As people become less exposed to the kind of thinking that offers truly momentous choices, they lose the ability to connect to it even during those few times when they are exposed. They look at a cornfield and think it is as complicated as things can possibly get.

This suggests that there is an authenticity inherent in unmediated cultural and biological evolution that is missing from global pop culture and biotechnology. Which is not to say that all differences are destined to disappear; "global monoculture" is just a metaphor. But, in comparison with what we have inherited, the differences that remain or come forth will be contemptibly narrow in scope. Moreover, if people are not made aware of the value of what we are destroying, they will be deluded into thinking this pauperized version of the world is equivalent to the real thing—"real" here meaning that which evolved without conscious human direction, and certainly without pecuniary considerations. The diversity that characterizes our inherited biocultural presence is what I call "authentic," and it is authentic in the same way that an original van Gogh is vastly more valuable than a copy regardless of the latter's quality (and this is not to deny that copies have some value) and *infinitely* more valuable than a fake. Similarly, we might someday be able to clone passenger pigeons from genetic material extracted from museum specimens and re-establish the species in the wild. I would be the last to argue that such a cloned species would be bereft of value, but nonetheless these Lazarus birds would never requite creation for the loss of the real thing. The momentousness of extinction is its irreversibility. Even if genetic engineering gives us the power to resuscitate extinct species, the authenticity and continuity embedded in the lineage will have been irredeemably destroyed.[48]

But isn't this whole line of reasoning based on a false premise? After all, isn't humankind's cultural–psychological space unlimited? People seem to have an endless capacity to innovate and make distinctions, no matter the richness or paucity of objective materials available to work with. Yes, the skeptics will admit, for the most part biological evolution grinds very slowly, and everyone agrees that the complement of Earth's species cannot be quickly regenerated after a mass extinction—in fact, it could take millions of years. But everyone knows that cultural evolution operates much faster. Want proof? Log on to the Internet. So what's the problem?

Such claims are frequently made and often accepted uncritically. They are partly based on the perfectly valid assumption that cultures have, in theory, a virtually unlimited capacity to proliferate. What they ignore, however, are fundamental limitations of time and space as expressed in the human genome. Each individual has a finite amount of time available for reflection. In 1990, the average life expectancy at birth for the world's population was 65 years (World Resources Institute et al. 1998, 2). Roughly a third of this span, call it 22 years, is spent sleeping. Let us arbitrarily further subtract another 3 years for infancy and senility, leaving a nice, round number of 40 years as the maximum amount available to the average human for critical thinking. Now, certainly none of us spends anything close to this amount of time engaged in innovation, fine discrimination, or any other kind of higher order thinking or cultural activity. Rather, our time is taken up with all manner of mundane activities. On the surface it may not seem as though the average consumer in a rich country spends much, if any, of his or her time doing things directly related to staying alive, but of course the whole system of wage labor is a surrogate for survival activities. And social organization dictates that the amount of time available for critical thinking varies greatly from person to person: philosophy professors have a lot, futures traders only a little.

Even if we were freed from all mundane demands and wished to spend all our time in reflection, the capacity of the brain to assimilate new concepts and bits of information into long-term memory is not unlimited. It has been estimated that it takes a minimum of 5–10 seconds of neuronal processing time to absorb a single new piece of information. That may not sound like much, but on an evolutionary timescale it (along with the small capacity of short-term memory) represents a real cost to the species in terms of time wasted and opportunities lost. There is, in short, an upper bound to how much humans can learn (Lumsden and Wilson 1981, 62–63, 337).

As further evidence, consider that we forget things. Why should that be? One might expect that we would have evolved perfect memories by now. The answer, of course, is that the physical capacity to remember (which is the material heart of our metaphorical cultural–psychological space) is only so big. James long ago recognized that the only way we can recall anything at all is by forgetting most of what we have learned. He quoted the psychologist Théodule-Armand Ribot to the same effect:

> As fast as the present enters into the past, our states of consciousness disappear and are obliterated. Passed in review at a few days' distance, nothing or little of them remains: most of them have made shipwreck in that great nonentity from which they never more will emerge. . . . If, in order to reach a distant reminiscence, we had to go through the entire series of terms which separate it from our present selves, memory would become impossible on account of the length of the operation. We thus reach this paradoxical result: that one condition of remembering is that we should forget. Without totally forgetting a prodigious number of states of consciousness, and momentarily forgetting a large number, we could not remember at all. Oblivion, except in certain cases, is thus no malady of memory, but a condition of its health and life. (James 1890, 1:680–681)[49]

All these observations point toward the principle of "cognitive economy," as psychologists now call it. It is simply the tendency of consciousness toward a minimum of complication wherever and whenever possible. We have already encountered this in our discussion of people's need for monothetic classifications, and to some extent the entire process of distilling sameness from diversity is also its kin. What remains to be emphasized, in order to dispel any notion that cognitive economy is merely some kind of hard-wired instinct for simplicity, is that humans discard on purpose. A goal is in view; and, ironically, expertise in achieving certain goals is enhanced by cognitive economy, not hindered by it:

> We grow unconscious of every feeling which is useless as a sign to lead us to our ends, and where one sign will suffice others drop out, and that one remains, to work alone. We observe this in the whole history of sense-perception, and in the acquisition of every art. . . . The marksman ends by thinking only of the exact position of the goal, the singer only of the perfect sound, the balancer only of the point of the pole whose oscillations he must counteract. The associated mechanism has become so perfect in all these persons that each variation in the thought of the end is functionally correlated with the one movement fitted to bring the latter about. Whilst they were tyros [novices], they thought of their means as well as their end: the marksman of the position of his gun or bow, or the weight of the stone; the pianist of the visible position of the note on the keyboard; the singer of his throat or breathing; the balancer of his feet on the rope, or his chin under the pole. But little by little they succeeded in dropping all this supernumerary consciousness, and they became secure in their movements exactly in proportion as they did so. (James 1890, 2:496–497)[50]

Looked at on an evolutionary time-scale, cognitive economy has enabled our species to effectively husband our cultural–psychological space, accounting for all its attendant limitations. What people have done with difference, and continue to do, is to use it as a wellspring for biocultural evolution.

OUR EVOLUTIONARY INHERITANCE

When seen aright, Earth's biocultural presence, the great fecundity of nature and culture, is the greatest asset humankind could ever have hoped for. It is a fortunate asset, not a birthright, for things could have turned out otherwise. Evolution, whether in nature or culture, is not programmed for ever-increasing variety; it isn't programmable for anything.[51] It is true that the tendency of evolution as it has played itself out *to this particular point in time* has been toward more life and more life forms and greater cultural variety—a fact with ethical implications, as I argue in the final chapter. But the five extinction pulses discussed in Chapter 2 show that evolution can contract as well as expand, even in the absence of human intervention. There is nothing inherent in the process of evolutionary change that is directing it toward more or less variety at any given moment. Rather than the richly varied natural and cultural fabric that we take for granted as inevitable, it is perfectly conceivable that life on Earth could have evolved so as to present us with conditions much closer to those that, say, a prisoner feels in solitary confinement. Rather than a world supplied with millions of species, thousands of languages and other cultural distinctions, and a tremendously varied landscape, we might have drawn one far more barren. We could have been born into a world populated by starlings and weeds, where every person spoke and dressed and ate and behaved more or less the same, where every field and town looked pretty much like any other. In such a world, the sun might seem to us as nothing more than a naked light bulb in a cell, each succeeding day offering the same small set of possibilities, our thoughts and aspirations severely circumscribed, our lives utterly routine, our consciousness without any inkling that a far richer diversity could (under other circumstances) exist.

But we were lucky. We got the world that we have. The one we have inherited is truly, even yet, a world of difference. At its heart is a paradox: *Human beings* need sameness, but *being human* means we first need genuinely rich stores of biocultural diversity to distill it from. If we continue to act in ways that destroy diversity, life of a sort will go on, but our aliveness—our uniquely human feeling of what life is supposed to be about—will have become extinct.

5

DIVERSITY AND THE HUMAN IDENTITY

We began this book by turning the tables on one of the best eavesdroppers of all time, listening in, as it were, on James Boswell as he tested his mettle against Voltaire's. Their exchange on that December day in 1764 culminated with Voltaire's insistence—perhaps serious, perhaps sarcastic—that the advent of a single great civic religion would make all mankind brethren. This served as a vehicle to introduce our main issue: the discovery of the meaning and value of diversity. It was an equivocal beginning, and perhaps the answer seems no clearer now. But there was no straight line to the truth. We have instead had to pursue the question over rough philosophical ground, seeking the counsel of evolutionary theorists, trying to gauge current global trends, making side excursions into specialized fields of inquiry. What we have arrived at is not a synthesis, not even a finished portrait, but at least the pursuit itself has been true to form, reflecting the deep ambiguity inherent in the subject matter.

Maybe, a skeptic might say, there isn't anything else *to* find. Maybe a collection of disparate facts and impressions is all there is. Maybe all these philosophical issues of sameness and diversity are just a delusion. It is a serious charge. As long ago as Plato's day the question of the one and the many could be made to seem like old hat. In a playful passage in his dialogue *Philebus*, Plato has Socrates say that "everybody has by this time agreed to dismiss as child-ish and detrimental to the true course of thought" all the silly questions about whether a being has a single essence or many distinct ones. Very quickly this line of reasoning bottoms out, landing on the flat proposition that everything is both one *and* many. Socrates goes on to tweak the student-philosopher who, in the first flush of discovery, finds that proposition exhilarating:

> Any young man, when he first tastes these subtleties, is delighted, and fancies that he has found a treasure of wisdom; in the first enthusiasm of his joy he leaves no stone, or rather no thought unturned, now rolling up the many into the one, and kneading them together, now unfolding and dividing them; he puzzles himself first and above all, and then he proceeds to puzzle his neighbours, whether they are older or younger, or of his own age—that makes no difference; neither father nor mother does he spare; no human being who has ears is safe from him, hardly even his dog, and a barbarian would have no chance of escaping him, if an interpreter could only be found. (Jowett 1952, 611)

One of his pupils, Protarchus, tosses a comeback line: "Considering, Socrates, how many we are, and that all of us are young men, is there not a danger that we and Philebus may all set upon you, if you abuse us?" But Socrates was kidding them in order to make a serious point, so he qualifies his joke, telling his listeners that there is a surefire way to cut through the confu-sion: classify, classify. The problem, says Socrates, is not that finding plurality in unity, and vice versa, is wrong; in fact, it is all too true. Nor is the problem the beguiling nature of the discovery, for the young scholar will eventually out-grow his initial infatuation. The real problem is that, of the people who pro-fess to be interested in the question of the one and the many, so few are will-ing to take the trouble to actually divide the entity in question into a definite number of constituents by means of a reasoned system. The "wise men of our time are either too quick or too slow in conceiving plurality in unity. Having no method, they make their one and many anyhow, and from unity pass at once to infinity; the intermediate steps never occur to them." Socrates scorns those lazy analysts who blithely jump from singularity to limitlessness without having specified what's going on in between: better they should keep silent "until the entire number of the species intermediate between unity and infinity has been discovered—then, and not till then, we may rest from division, and without further troubling ourselves about the endless individuals may allow them to drop into infinity" (Jowett 1952, 612). This is Beckner's Proviso all over again. Without thoroughness of method in classification, the question of the one and the many is indeed just so much hairsplitting. Socrates concludes: "The infinity of kinds and the infinity of individuals which there is in each of them, when not classified, creates in every one of us a state of infinite ignorance; and he who never looks for number in anything, will not himself be looked for in the number of famous men" (Jowett 1952, 612).[1]

William James was still musing on the same subject more than two millennia later, still thinking it worthy of serious study, still finding no easy answers:

These profundities of inconceivability, and many others like them, arise from the vain attempt to reconvert the manifold into which our conception has resolved things, back into the continuum out of which it came. The "many" is not the concept "one"; therefore the manyness-in-oneness which perception offers is impossible to construe intellectually. Youthful readers will find such difficulties too whimsical to be taken seriously; but since the days of the Greek sophists these dialectic puzzles have lain beneath the surface of our thinking like the shoals and snags of the Mississippi River; and the more intellectually conscientious the thinkers have been, the less they have allowed themselves to disregard them. (James 1911b, 91)

It seems, then, that sameness and diversity are subjects of perennial interest to reflective minds. Today, the issue is often debated in terms of the fragmentation of the intellectual landscape, which, according to one's outlook, is either a worrisome sign of decay or a welcome corrective against hubris. Erosion, of a kind, is responsible for the dominant feature of the terrain: a chasm between science and the humanities, now grown so wide and deep that it is often given up as unbridgeable.[2] Actually, "given up" is a mild way of putting it. There are plenty of people who positively relish the distance, thankful of any opportunity to dismiss the other side, on guard always against any attempts at bridge building.

This imaginary landscape is a caricature, but, like all caricatures, the exaggerated details serve to highlight a core truth. Anyone who takes an interdisciplinary view of biological and cultural diversity must deal with the chasm.[3] As a matter of procedure, then, it would seem pretty risky to postulate any kind of continuity (or even affinity) between the two great realms of difference we live with. But, apart from a desire to bridge intellectual rifts, why should anyone really care one way or the other? Because by seeking a holistic understanding of diversity, we gain a more accurate picture of how each of us, as individuals, shares in the collective life of humankind. If there is such a thing as "our common humanity," as Voltaire implied there could and should be, we would do well to examine what it is made of. It is a question of the first importance, for "we cannot be unaware . . . of the extremes—called 'apartheid' and 'final solution,' among other things—to which giving up the ideal of the unity of the human race can lead" (Todorov 1993, 88). The paradox is that we can only grasp what is universal by first recognizing what is different. To do the reverse runs us down the blind alley of ethnocentrism, "the unwarranted establishing of the specific values of one's own society as universal values" (Todorov 1993, 1). (It is not difficult to see that a similar fallacy has helped create the schism between science and the humanities.) Every person, from the tribal member in New Guinea to the most case-hardened New Yorker, has personally to arrange the natural and cultural facets of existence. Surely, our differing responses to nature and culture, to the variety they embody, must be accounted for in the search for commonalities.

ARE BIOLOGICAL AND CULTURAL DIVERSITY REALLY RELATED?

We should never lose sight of the fact that diversity is an organizing principle. To assess diversity is to make judgments about *how much* such entities as species and languages differ from each other, not merely to determine that they are different, and this implies a system to classify the differences. If making such distinctions is important, it is important first of all because it helps us see the world in a more accurate light. "Numbers alone do not make science; it is relations between numbers that are needed," wrote the biologist Robert MacArthur. "Applying a formula and calculating a 'species diversity' from a census does not reveal very much; only by relating this diversity to something else—something about the environment perhaps—does it become science. Hence there is no intrinsic virtue in any particular diversity measure except insofar as it leads to clear relations" (MacArthur 1972, 197).

One can make a strong case that this illuminative power of diversity is what is paramount.[4] To grasp the magnitude and variety of biological and cultural wealth around us is to gain a better perspective on our relationship to the rest of life. To be concerned about their loss is to better understand our responsibilities to other species and to the rest of our own. For biologists and linguists working to stop extinctions, diversity is a way to organize and value the world.

If, in turn, average people are to value diversity, they must be able to grasp it intuitively. Species and languages, for all their faults and ambiguities, are the concepts that currently work best to measure diversity. A powerful argument for preserving them is that they make the planet a better place to be. The case for diversity must not fail to appeal to this value. "We do not understand our- selves yet and descend farther from heaven's air if we forget how much the natural world means to us," writes Edward O. Wilson. "Signals abound that the loss of life's diversity endangers not just the body but the spirit." We must strive, he says, for an "enduring environmental ethic" that aims "to preserve not only the health and freedom of our species, but access to the world in which the human spirit was born" (Wilson 1992, 351).

Even if all this is granted, it has to be admitted that an intellectual and moral program promoting kinship between biological and cultural diversity will always strike some people as hopelessly diffuse and therefore pointless—or worse. Such is the charge leveled by the American conservation biologists Reed F. Noss and Allen Y. Cooperrider:

> On the face of it, inclusion of social diversity in a definition of biodiversity makes sense. We are fundamentally as much a part of Nature as any other species and share kinship and ecological interactions with all of life. But what would be the practical effect of including diversity of human languages, religious beliefs, behaviors, land management practices, etc., in a biodiversity definition and striving to promote this diversity in conservation strategy? We believe the effect would be to trivialize the concept and make it unworkable, even dangerous. As Kent Redford (personal com- munication) notes, "This definition allows Manhattan or Sao Paulo to be considered on equal footing with the Great Barrier Reef of Australia and makes impossible any coherent discussion of biodiversity conservation." We are not interested in main- taining social and cultural diversity if it means maintaining Nazis, slave owners, or those who enjoy using desert tortoises for target practice. This book is about how culture might adapt to nature. We want to conserve all cultural approaches that are compatible with conserving biodiversity. To combine cultural and biological diver- sity into one definition is to muddle the concept. (Noss and Cooperrider 1994, 14)

The authors oppose conflating cultural and biological diversity, yet (as they recognize), on a strictly logical basis, that objection falls apart when one remembers that culture is, at bottom, behavior. If one considers human cul- tural diversity to include the entire range of behaviors of various populations of the primate *Homo sapiens*—a perfectly reasonable, perhaps even requisite, standpoint for a biologist to adopt—then cultural diversity becomes a matter for ethology, the study of animal behavior. It may be possible to come up with a definition of biodiversity that excludes the facts and consequences of ethology, but I doubt if many biologists would find it satisfactory. Noss and Cooperrider accept that humans do not exist apart from or above the natural world, and surely they believe that humans are the product of evolution by natural selection. On these grounds it is illogical to treat the variety of our behaviors any differently from those of other species, leaving us to conclude that cultural diversity is undeniably a part of biological diversity.

But Noss and Cooperrider quite rightly zero in on the *practical effects* on conservation of combining the two. Perhaps there are good strategic reasons for keeping cultural and biological diversity separate. Here, the authors are worried that combining them would play down the gravity of biodiversity conservation and make it unworkable, or, somehow, even dangerous. The thrust of their thinking seems to be that embracing cultural diversity means abjuring all value judgments, that it means accepting any and all beliefs and practices (such as Nazism and slave-holding). Consider the trenchant comparison between Manhattan and the Great Barrier Reef. The cut of this remark by Redford (also a conservation biologist) is the fear that scientific efforts[5] to preserve biodiversity will be diluted by propounding cultural diversity, that by embracing the latter conservation biologists would be compelled to consider a city as equivalent in value to a richly diverse natural ecosystem. Noss and Cooperrider do not want any part of relinquishing their right to make value judgments. They want to promote their own ethical values and eliminate opposing ones.

Well, so do I.

Now, most people could easily rattle off dozens, probably hundreds, of specific cultural practices that they find reprehensible. My list would range from the so-called honor killings in some parts of the Middle East; to the slaughter of endangered species to satisfy demands for traditional Asian medicinals; to the selfish propensity of Americans to drive absurdly oversized, wasteful cars, thereby depleting the world's fossil fuels and polluting the planet beyond all reasonable necessity. I can understand, if perhaps only at a basic level, the various reasons that are given to justify each of these cultural practices. Furthermore, I can appreciate that some of them are products of cultures other than my own and that I therefore may never be able to *fully* understand why others value them. So I acknowledge the autonomy of other cultures and the beliefs contained in them. Yet despite my support for preserving cultural diversity, I still wish to see these particular cultural practices disappear. I take this stance unapologetically, because it really is a fallacy to say that one cannot be an advocate for cultural diversity just because it encompasses snuff films as well as the Sistine Chapel.

AVOIDING THE TRAP OF RELATIVISM: POLYTHETIC MORALITY

I am criticizing Noss and Cooperrider's quotation not to pillory the authors, for I share the high regard in which their work is held within the field of conservation biology. The point is that they have clearly enunciated a widely held assumption: that advocating for cultural diversity is the same as advocating a flaccid cultural relativism,[6] a doctrine that "is as indefensible at the level of logical consistency as it is at the level of contents":

> The relativist inevitably ends up contradicting himself, since he presents his doctrine as absolute truth, and thus by his very gesture undermines what he is in the process of asserting. Furthermore and more seriously, the consistent relativist writes off the unity of the human species. . . . The absence of unity allows exclusion, which can lead to extermination. What is more, a relativist, even a moderate one, cannot denounce any injustice, any violence, that may happen to be a part of some tradition other than his own: clitoridectomy would not warrant condemnation, nor would even human sacrifice. Yet it might be argued that the concentration camps themselves belonged, at a given moment of Russian or of German history, to the national tradition. (Todorov 1993, 389)

What we must do, according to Todorov, is simply acknowledge that we humans all belong to the same species: "that is not very much, but it is enough to serve as a basis for our judgments," allowing us to recognize that "justice is nothing but another name for taking the entire human race into account." This "founding principle of ethics" is augmented by a political principle the French philosopher Baron de Montesquieu stated long ago: that "the unity of the human race must be recognized, but also the heterogeneity of the social body. It then becomes possible to make value judgments that transcend the frontiers of the country where one was born: tyranny and totalitarianism are bad in all circumstances, as is the enslavement of men or women. This does not mean that one culture is declared a priori superior to others, a unique incarnation of the universal; but it does mean that existing cultures can be compared, and that more may be found to praise in one place, more to criticize in another" (Todorov 1993, 390–391).

> Let us break down simplistic associations: demanding equality as the right of all human beings does not in any way imply renouncing the hierarchy of values; cherishing the autonomy and freedom of individuals does not oblige us to repudiate all solidarity; the recognition of a public morality does not inevitably entail a regression to the time of religious intolerance and the Inquisition; nor does the search for contact with nature necessarily take us back to the Stone Age. (Todorov 1993, 399)[7]

Todorov's declaration rings out with good sense. But can we affirm it while at the same time make an ethical case for the preservation of biocultural diversity per se?

If we look backwards for guidance within the ethical tradition of the West, the initial results are not promising. To some extent, the whole history of Western moral philosophy has been a search for a universally applicable formula governing "right" values (and the behaviors that ought to issue from them). In other words, it has been a search for a monothetic consensus on core values. Anyone who has dared suggest that this might be a waste of time, a case of looking for the wrong thing, has been immediately cast out and branded an anarchist, a nihilist, or, at best, a relativist. Yet the plain fact is that people do not agree on ethics, that the actual state of affairs is one of moral pluralism. People *do* use desert tortoises for target practice, and, much as may I despise them for it, they think there's nothing wrong about it. The only way an absolutist approach to ethics can deal with this is by stubbornly and repeatedly asserting that one's own code is right and everybody else's is wrong. So far it has been a spectacularly ineffective approach, for it is evident that all the religious and ideological argumentation since the dawn of history has as yet done little to convince people to adopt a single code of conduct—to adopt Voltaire's solution, as we have called it. It may yet happen (not quite literally, perhaps, but practically), because globalization is now creating, for the first time, the technological and social conditions necessary to achieve widespread homogeneity, in ethics as well as other aspects of culture. Whether such homogenization actually comes about remains to be seen. In any case, I have tried to argue that the threat is real.

What I wish to emphasize here is the special quality ethical assertions must take on when dealing with diversity as a condition to be valued. To say "diversity is good and ought to be valued" is radically unlike most traditional value statements, which propose that some *specific* characteristic or behavior (e.g., truthfulness, charity, compassion, bravery, etc.) ought to be valued. When I say "diversity ought to be valued," what I am really saying, in terms of ethics, is that we ought to embrace the fact that there are different, sometimes conflicting, moral norms—that moral pluralism as such is a good thing, better than Voltaire's solution even if it could be shown that attainment of universal ethical agreement would result in less strife among the world's people. The reason is because the distillation of sameness from diversity is essential to whatever it is that makes us human, no less in ethics than in any other sphere. A world where everybody held the same values might indeed be far more peaceful than the one we have now, but it would not be a human world in any recognizable sense.

This stance in favor of moral pluralism does *not* commit one to relativism, as is so often charged. There is another, nonrelativistic alternative to absolutist ethics, one that I shall call "polythetic morality." A polythetic morality is based on a set of two linked assertions. First, that there is great value in the failure of *all* people to share *all* core values of morality absolutely; that is, it is good that a monothetic consensus on values does not exist. Second, that there is equally great disvalue in the failure of *most* people to share *most* core values pluralistically; that is, it is bad that a polythetic consensus on values does not exist. Unlike absolutist ethics, a polythetic morality is satisfied that not everyone agrees on every moral issue. Unlike relativist ethics, a polythetic morality does not disown assertions of right and wrong and other value judgments; it recognizes that competing moral assertions (including absolutist ones) are part of a larger ethical landscape whose overall diversity is worth being preserved. Under a polythetic morality, people strive to convince others of their moral code, and it is this process of striving that is good. If, however, any one moral code should triumph absolutely, it would be bad. The reason is that people are often simply mistaken about right and wrong, good and evil. No matter how rational, enlightened, and progressive a given ethical stance may appear to its proponents at a particular moment in history, any kind of moral certitude closes down self-criticism. At worst, it leaves societies susceptible to demagoguery, to twisted dreams of apartheid, master races, and final solutions. At best, it stunts the capacity for ethical reflection and leaves people less able to adjust their behaviors to new social circumstances.

How *can* human unity be achieved without uniformity? Long ago Rousseau suggested an answer. According to him, "the principal characteristic of human behavior is *perfectibility*—that is, something that has no positive content but that allows human beings to acquire all contents." Thus universality gets "an unexpected twist: what is common to human beings is not any particular trait," but rather their freedom, "their capacity to transform themselves"—potentially for the better (Todorov 1993, 21–22). So if we define the achievement of unity among humankind as the best state of life, and if, with Rousseau, we accept freedom as our behavioral imperative, we see then that we cannot achieve unity through uniformity because uniformity throttles freedom. And, deprived of freedom, we would be denied that which lies at the heart of our behavior, that which is truly common to us all. The paradox is that perfectibility, rather than perfection, is what must always lie ahead of us; the striving, not the getting. The freedom to improve ourselves can come only from a world of difference, and the wider the spectrum of genuine diversity, the more the scope of freedom expands. But if we allow ourselves to be lured by uniformity, by imagining we will be just as happy, happier even, in a world where everyone speaks the same language and sees the same weeds, then we will have lost both the hope and the means of perfectibility. The striving will be over.[8]

James A. Nash has proposed a form of striving as the basis for an ethic that takes in all species, not just humans. "[T]he one necessary criterion for the recognition of biotic rights that I find compelling is *conation*—that is, a *striving to be and to do*, characterized by aims *or* drives, goals *or* urges, purposes *or* impulses, whether conscious or non-conscious, sentient or non-sentient. At this point, organisms can be described as having 'vital interests'—that is, needs or goals—*for their own sakes*. These conative interests provide a necessary and minimally efficient status for at least elementary moral claims against humans" (Nash 1999, 472). Conation covers all living things, thus excluding abiotic components of the biosphere, though, as Nash is quick to note, because ecosystems depend on the latter and humans depend on ecosystems, we must therefore treat the entire biosphere with respect.

> Grounded in conation, biotic rights apply to both individuals and species, because individuals and species seem to be constitutive of one another—inseparable and interdependent. Even if a species as a genetic lifeline from the past for the future is not conative in itself (which I doubt), that claim would not contradict a theory of biotic rights grounded in conation. The reason is that a species as a genetic lifeline

is not only the aggregation of conation in present populations but also the carrier of conation for all future generations. So, we can argue that recognizing the rights of species is essentially the same as recognizing the rights of future generations. Species can be said to have at least *anticipatory rights* in the sense that we can reasonably expect that they will exist and have vital interests, unless we deprive them of that potential. Therefore, the human community has anticipatory obligations to preserve otherkind's conation for the future. (Nash 1999, 473)

This is an important extension of volition as couched in the Jamesian tradition of free will. Lack of sentience no longer excludes beings from moral consideration. Of course, animal rights activists have made the same claim, but Nash's emphasis on maintaining continuity in the flow of life, of evolution, allows us to escape the noose of absolutist ethics, whatever brand it may be. He is careful to say, for example, that his proposed biotic rights "are not moral absolutes" but can be overridden for "just cause," "such as self-defense against a pathogen, the satisfaction of basic human needs, or even the culling of alien species to protect an ecosystem" (Nash 1999, 472). Thus, unlike animal rights advocates, Nash is willing to accept that a certain amount of anthropogenic killing may be necessary or even desirable to maintain the overall flow of life. All the same, however, I suspect he would agree with me that we ought to minimize the amount of killing that needs to be done; if so, it is a sensible position, and accords with the precepts of polythetic morality.

To achieve a polythetic consensus on core values would be to achieve unity in the truest sense because such broad (but not absolute) agreement confirms the psychological mechanism upon which our common humanity is founded. Uniformity is often confused with unity, but the basic difference between them is this: To realize uniformity is to achieve an end-point, whereas to realize unity is to engage in a continuous, never-to-be-completed endeavor. In line with Jamesian pluralism, there is always "more" to the realization of unity, always new and changing conditions to take account of. This is also in line with Beckner's Proviso: One must have a wide knowledge of different cultural approaches, and a commensurately wide knowledge of how nature functions, before one can begin to define a meaningful set of core values on which to found a polythetic consensus. The corollary is that the realization of unity demands the continuation of the odyssey of evolution, and therefore requires an ongoing biocultural presence. Thus, in defining core values, one must give serious consideration to the ethical claims of other species, as well as to the need to preserve the diversity of our interactions with them. Nothing less is required for a fully polythetic morality.

The price one pays for adopting polythetic morality is that one must accept that no absolute ethic can ever preserve diversity—including one that says we must preserve diversity absolutely. The fact that extinction has happened since the beginning of life (Heywood 1995, 208) is important, though not because of the erroneous inference that extinction is inevitable and so we needn't be concerned about it. Rather, it shows that there is no precedent in evolutionary history from which to derive an ethic that says *all* extinctions should be prevented. So the quest is not to preserve every product of diversity, but to preserve the viability of the evolutionary processes that produce diversity. Note, however, that preserving the processes will necessitate preserving a large portion—in fact a very large portion—of the products.

Now, any fool can see that "Preserve process!" is a nonstarter as a call-to-arms. People will always focus on the panda, not the panda's phylogeny. This is where Nash's idea of biotic rights can inform the biodiversity concept. Combining the two allows us to acknowledge the *totality* of the products of evolution in a satisfying way, to affirm the importance of all species while not compelling us to hug every last one of them. It would even sanction us, under the limited circumstances outlined by Nash, to occasionally destroy a species, such as the smallpox virus, whose existence is so detrimental to human interests that absolutist objections to species destruction ought to be overridden.[9] Most importantly, it would give scope to the preservation of *all* the processes that produced the biodiversity because there is no other way to preserve the totality. Safeguarding biodiversity is therefore an ethic on a global scale: It will not do to protect it in North America if it is devastated in Africa. That would be as complete a failure as if it had been destroyed on both continents.

Thus it makes sense, and is only a seeming paradox, to pursue the preservation of biodiversity *in toto* while knowing full well that the goal is impossible, that at least some ongoing amount of extinction is inevitable—and, in a tiny number of cases, desirable. If Nash's biotic rights principle were properly and judiciously applied, the number of such cases would indeed be very few: because we still know so little about the processes that produce biodiversity, it is incumbent upon us to use precaution when dealing with the products. Speaking practically, as far as I can see the class of species so inimical to humans that elimination is *possibly* warranted is limited to infectious diseases. The AIDS virus, yes; rattlesnakes, dust mites, poison ivy, and cockroaches, no.

The same approach can be applied to cultural diversity, but there is one basic difference. Species are the main carriers of biological evolution, languages the main carriers of cultural evolution. In both cases the guiding principle is preserve the diversity of the carriers. However, whereas we have just deemed it possible to justify the elimination of a small number of a certain class of species, it is always a mistake to destroy a language, the basis for entire cultures. I cannot conceive of it ever being right to for people outside of a given speech community to directly eliminate that community's language (or, in what comes to the same thing, create conditions that coerce the speech community into giving up its language), because no language is inherently inimical to human interests in general.[10] Nonetheless, if the members of a speech community *freely decide* that it is in their best interest to give up their language in favor of another, that is their right; so in this limited sense there is no warrant to preserve every language absolutely. But, as with species, the number of instances where this will happen should, I suspect, be vanishingly small. The overwhelming evidence now emerging from language preservation activism is that most speech communities want to maintain their languages if at all possible. The right to self-determination in language use is an assertion of the right to maintain the vehicle of one's culture, and is fundamentally different from blanket assertions of cultural autonomy in which it is held that every cultural practice is sacrosanct and cannot be criticized by outsiders. All of us can and ought to work for the elimination of specific cultural practices we find reprehensible—remembering, however, that others with whom we disagree have an equal right to promote their point of view, and that it would be best for humankind if any one set of views, including those we cherish most, didn't carry the day absolutely.

What we have just done—joining polythetic morality, Nash's concept of biotic rights with its emphasis on conation, and a biocultural approach to the preservation of diversity—amounts to a redefinition of the principle of plenitude we discussed in Chapter 1. As you recall, it held that "the extent and abundance of the creation must be as great as the possibility of existence" and commensurate with the perfection of God, so that "the world is the better, the more things it contains" (Lovejoy 1936, 52). In other words, every difference that can be conceived to exist *must* necessarily exist, so that the universe is like a plenum exhaustively filled with examples of diversity. In contrast, the new principle of plenitude is polythetic, not absolutist: it enjoins us with Beckner's Proviso, which here translates into the precautionary requirement that it is best to know a great deal about something before we commit to irreversible actions regarding it. The new principle of plenitude is at once more modest and not so expansive as the old, yet far more realistic, when it states: "The world is the better for the diversity it contains."

THE COMING THRESHOLD

The examples presented in this book begin to show that biocultural variation is discontinuous, with diversity described by modal points of identity.[11] I believe that the evidence, on balance, shows that the real world cannot be reduced to a collection of monothetic "either/or" categories. Rather, it is as William James thought it to be, a concatenation of partially overlapping categories—a polythetic universe. In this universe of ours there is an inherent affinity between biological and cultural diversity. But whether or not one accepts this conclusion on theoretical grounds, it is hard to ignore the similarities between the practical forces driving biological extinctions and cultural homogenization. And it is in this real-world realm that a genuine sense of crisis has taken hold, quite independent of theoretical considerations. The feeling of crisis is driven by the conviction that we soon will reach a perilous threshold of loss, a Rubicon of extinction, a point of no return beyond which a critical amount of biological and cultural diversity will have been destroyed, never to be regenerated on any time scale significant to the development of humankind.

Without such a long-range, encompassing view, touting a converging diversity crisis becomes merely polemical. Recalling Chapter 2, consider where the world stands today in relation to the course of life over the ages. Though proportionally most species that have ever existed are now extinct (to the nearest order of magnitude, so runs the joke, *all* species are extinct), the momentum of evolution down to the present day has been toward a fecundity of life forms. Taxonomy is still a viable career choice because we have millions of species, not thousands or hundreds. Likewise, the variety of human expression and organization still extant is, by any standard, little short of astonishing. To reiterate, there are, depending on how one counts them, something on the order of 6,800 oral languages still spoken as mother tongues, more than 4,000 "distinct cultures known to anthropology" (as cited in Durham 1990, 194), some 10,000 religions, an untold number of artistic forms, kinship systems, and so forth—at least tens of thousands of cultural differentiae in all. The biological and cultural diversity now existing is the pre-eminent fact of life, a first-order condition of Earth's being. Nothing less than this biocultural presence is now at stake. The essential nature of diversity combined with a looming threshold of irreversible diminishment justify the label "crisis."

The prospect is taking shape against a new backdrop: a rapidly emerging global culture fueled by the telecommunications revolution, the near-collapse of Communism and the unbridled diffusion of (supposedly) market-based ideologies, the unprecedented reach and political influence of transnational corporations, the primacy accorded Western scientific and technical knowledge, the availability of worldwide travel, and on and on. Globalization affects biodiversity in many ways: by easing the spread of exotic species, making the logging of remote rain forests profitable, undercutting proven indigenous land-management systems. For nature, the problem is the speed of change as much as its extent:

> Abrupt changes, changes with no parallels in normal biological or climatological processes, scramble the feedback messages and leave the system in disarray.[12] Ecosystems have few mechanisms for adapting to expanses of concrete, and these act too slowly to protect the system paved over from simplification and collapse. . . . The process of recovery is slow and incomplete; opportunists take and hold possession for a longer time, making it more difficult for full regeneration to take place. Thus, the ontology and epistemology of the ecological world view give rise to a positive value—that of harmony with nature and nature's way. It is good, in this view, to do things in a way that mimics nature's patterns; it is good to promote the natural processes that, if not interrupted, produce greater diversity; it is good to introduce alterations slowly enough to allow nature to react. And it is bad to thwart those natural processes, to interrupt well-established patterns, to introduce irreversible changes. (Norton 1987, 207)

The same ethical principles apply to cultures: It is better for change to take place on a scale and at a pace that can be intelligibly assimilated by people, whether individually or collectively. However, we seem headed for a future where, in Benjamin Barber's memorable phrase, "velocity is becoming an identity all its own" (Barber 1995, 194). Interchange between cultures is not bad, but, as many critics of globalization have pointed out, what we have now is not interchange, but the flooding of the world with Western (largely American) consumer culture. With the flow so one-sided, so fast, and so extensive, it is not surprising that American popular music drives out or hybridizes traditional folk music; that English displaces indigenous languages; that processed foods brought in on container ships replace traditional, locally produced fare in daily diets; and that proselytizing missionary work imposes monotheism in place of native spiritual beliefs. In some instances the change is innovative, to be sure. But much of it simply destroys less politically powerful forms of cultural expression and organization.

So the diversity crisis is real. Yet, based on current knowledge, it cannot be said to issue from the likelihood of a descent into what might be called "terminal scarcity." As stated in Chapter 2, no one knows how many species exist, but a figure of 5–15 million is considered reasonable, and most recent estimates of near-term extinction prospects put the impending global species loss at 1–10 percent per decade (Stork 1997, 62–65). If we extend any combination of these per-decade projections over the next century, we are still left with hundreds of thousands, probably millions, of species intact. Turning to the most accessible indicator of global cultural diversity, the status of the world's languages, we reach a similar conclusion. The possible extent of mother-tongue extinctions over the next century is very high, with first-generation guesstimates running from 20–25 percent to, in worst-case scenarios, 90 percent (Krauss 1992, 7). Yet no matter how one counts languages (as opposed to dialects), even a 90 percent reduction would still leave a minimum of several hundred intact (Harmon 1995, 17). Using these numbers, no one can say that we face the prospects of an absolutely depauperate planet or an actual global monoculture.

But it would be a grave mistake to take comfort from this or use it as an excuse for inaction. We have no idea how many species we can lose before essential ecosystem services (purification of water, mitigation of floods and droughts, pollination of plants, and many others discussed in Daily 1997) and overall environmental quality are seriously compromised. As Stephen R. Kellert rightly observes, "removing 10 percent or even 1 percent of the planet's species" would be akin to "randomly destroying pieces of an extremely complex mechanism while blindly hoping not to damage some vital element or process" (Kellert 1996, 31). Nor do we know how much cultural distinctiveness can be lost before we begin slipping toward a *de facto* totalitarianism. Alfred North Whitehead considered the "Gospel of Uniformity" almost as dangerous a threat to social progress as the "might makes right" doctrine of brute force:

> The differences between the nations and races of mankind are required to preserve the conditions under which higher development is possible. . . . A diversification

among human communities is essential for the provision of the incentive and material for the Odyssey of the human spirit. Other nations of different habits are not
enemies: they are godsends. Men require of their neighbors something sufficiently
akin to be understood, something sufficiently different to provoke attention, and
something great enough to command admiration. (Whitehead 1925, 297, 298)[13]

The question of biocultural extinctions is truly momentous because the
losses cannot be reversed. The criterion of irreversibility, sad to say, plays far
too small a role in conventional economic decision-making. Conservationists
have long urged the adoption of a simple "precautionary principle" in decisions
where significant, permanent environmental effects are anticipated. It says that
where their likely magnitude or potential reach is unknown, the action being
proposed should err on the side of restraint and caution as much as possible.
It is an eminently prudent strategy and one that ought to be unobjectionable.
But there are still plenty of "cornucopians" out there who pooh-pooh the evidence for biological extinctions (and believe that boundless human technological ingenuity and infinite resource substitutability will bail people out of all
environmental messes anyhow), as well as triumphalists who believe that cultural homogenization is inevitable (explaining, for example, that English has
come to dominate the world scene because of purely intrinsic reasons or by
sheer chance, rather than by its having been made the vehicle of political and
economic dominance). They have been all too successful: the onus is still on
proponents of diversity to prove that standard economic and political behavior will damage the biocultural presence. By demanding exhaustive, *absolute*
evidence of damages before conceding that action should be taken to stop the
destruction of diversity, entrenched powers effectively protect the status quo
and devalue the biocultural presence.

FROM UNIFORMITY TO UNITY

One last question remains: Does the existence of diversity itself, and the fact
that it is facing a crisis, carry any decisive ethical implications? Is there an ultimate moral imperative to preserve diversity?

Many people would say the answer is simple: Humans have an innate need
for diversity and therefore we must preserve it.[14] In light of James's conception
of consciousness, this well-intentioned reasoning does not go nearly far
enough. Diversity is not just one need on a par with dozens of others. If it is

the means through which our consciousness function operates, and if consciousness is what makes us human, then diversity makes us human. When we act in ways that reduce diversity, whether in the nonhuman world or in our own cultures, we corrode our essential humanity. The quality of humanity is achieved through the continual mental activity of evaluating differences and resolving them (albeit often provisionally) into individualities. This largely unconscious process of comparison, which we all carry out countless times each day, is, upon analysis, almost inexpressibly sophisticated. Whatever it means to "be human," this process cannot be omitted.[15] To the extent that its richness is impaired—as would happen if biological and cultural diversity were sharply curtailed—that much are we prevented from enacting the full potentiality of being human. James knew that diversity is the field against which we identify things and events as being "the same" and that, without our archetypal notion of sameness, deprived of the "psychological sense of identity" to deal with objective diversity—"the most important of all the features of our mental structure"—the human mind would be fundamentally changed (James 1890, 1:460). When the field of diversity shrinks, the potential for distilling sameness from difference is impoverished commensurably. If, at some point, the impoverishment becomes acute enough, then, as I interpret it, our species will have passed a threshold. We will have become something other than human.[16] By this view, then, diversity ought to be preserved because it provides the grounds for the continuance of *Homo sapiens* and the humanness that we recognize in ourselves.

But even this far-reaching rationale does not fully establish the value of continuing the biologically diverse systems that support all life on Earth. For the contemporary environmental philosopher Holmes Rolston III, any ethical treatment of species (including ours) has as its prerequisite the acceptance that they "be objectively there as living processes in the evolutionary ecosystem," for "if species do not exist except embedded in a theory in the minds of classifiers, it is hard to see how there can be duties to save them. No one proposes duties to genera, families, orders, or phyla; everyone concedes that these do not exist in nature" (Rolston 1985, 721).[17] Rolston takes the varied criteria for species (descent, reproductive isolation, morphology, gene pool) as evidence of their objective existence.

At this point, we can anticipate how there can be duties to species. What humans ought to respect are dynamic life forms preserved in historical lines, vital informational processes that persist genetically over millions of years, overleaping short-lived individuals. It is not *form* (species) as mere morphology, but the *formative* (speciating) process that humans ought to preserve, although the process cannot be preserved without its products. Neither should humans want to protect the labels they use, but the living process in the environment. (Rolston 1985, 722)

Though species in general, unlike *Homo sapiens* in particular, are not moral agents, Rolston defends their biological identity, declaring that "the dignity resides in the dynamic form; the individual inherits this, instantiates it, and passes it on" (Rolston 1985, 722).

This is not to say that the evolutionary process itself is teleological or invested with morality. Evolution by natural selection predates *Homo sapiens* by millions of years and will likely continue, barring a planetary cataclysm, after our species has disappeared. The process churns on, distilling new forms of life, heedless of human wishes, unguided by purpose on or beyond the earth. An explanation of its significance will fit on a bumper sticker: Speciation Happens. The production of new species is a warrant that nature really exists, that "The Big Outside" (in the happy phrase of environmental activists Dave Foreman and Howie Wolke) really is at once singular, large, and "out there" apart from subjective human perceptions.[18]

Nonetheless, as Rolston demonstrates, we can derive morality from the fact of evolution. I extend his reasoning to biological and cultural diversity as a whole. The moral imperative for preserving diversity is the continuance not just of humankind, but of the biocultural evolutionary process that produced us and every other species and brought us all to where we are—together.[19]

Cultures need a base of variety from which to work if they are to generate new differences and thereby avoid stagnation.[20] The diversity must be genuine, that is, it cannot rest on narrow pecuniary considerations. And, as noted above, change must unfold at a scale and pace that makes it capable of being assimilated by the men, women, and children of a particular culture. Such genuinely evolving cultural diversity is necessary "for the constant rehumanization of humanity in the face of materialism" (Fishman 1982, 6).[21]

Nor can we do without biodiversity. "Nature and I are two," wisecracked Woody Allen in one of his early movies—a great satiric line, part of his angst-ridden, urban-intellectual shtick.[22] Still, it *is* just a joke, even if many sociosatiated city-dwellers do manage to exist with minimal direct contact with nature.[23] For even in the deepest canyons of Wall Street, everybody eats and everybody breathes.[24] But, as I hope is clear by now, there is more to the connection than this. Kellert reminds us that "living diversity continues to serve as an essential medium for the developing person" because "the ultimate raw material for much of human industry, intellect, emotion, personality, and spirit originated—and remains rooted—in a healthy, abundant, and diverse biota" (Kellert 1996, 27, 31). No matter how we might wish to insulate ourselves from the vagaries of nature, "living diversity remains an essential element of human language, myth, and story, a vital source of our notions of beauty and understanding. The creatures of the world inspire and instruct. They nurture us intellectually and enrich us emotionally" (Kellert 1996, 32). Cultures different from our own perform these functions as well, and as such complement natural variety. "Knowledge of others," concludes Tzvetan Todorov, "is not simply one possible path toward self-knowledge: it is the only possible path" (Todorov 1993, 84).[25] Active affinities with nature and culture are thus complementary halves of a truly whole person. In the end, Kellert says, "the human species can no more dissociate itself from the natural world than it can divorce itself from the products of cultural creation" (Kellert 1996, 27).[26]

Kellert's point about diversity as a person-making force suggests, rather like the theodicies of the theologians, that conflict, discord, and trial are necessary parts of a truly human existence. Such explanations traditionally have been used to reconcile the existence of evil with a loving God.[27] Whether or not one finds them plausible or even necessary, they are an honest attempt to explain why peace and harmony have eluded us. I think those of us who preach that diversity is good need to be at least as honest as that. Let us, therefore, face squarely the cold fact: Valuing diversity in nature and culture means giving up all hope of achieving paradise on Earth.[28] Paradise as traditionally conceived, that is: a static, monolithic, perfectly composed, and forever-tranquil realm of bliss. That image of paradise (variants of which, incidentally, are heavily promoted by corporate advertising under the general guise of "You *can* have it all") is a chimera, forever beyond our reach. And we should thank our lucky stars that it is. To be human, we need to strive for this kind of paradise, but it would be the most inhuman act in the history of creation if we ever attained it.

Now, such a conclusion will strike many people as simply perverse. But my intent is not to champion discord, which I would like to avoid as much as anyone. What I do say is that the *process* of dealing with discord—whether that be avoiding it, confronting it, or overcoming it—has its own value as part of a much broader general process that we can call "endeavor." Thinking and doing, planning and regrouping, succeeding and failing: Engaging in all of these is what it means to be alive. To live is to endeavor. And, as it happens, the polythetic universe is the perfect setting for life—*all* life, as the redefined principle of plenitude would have it.

The ethical undercurrent throughout this book has been that we should value what has evolved over millennia and without conscious human direction, rather than what will likely emerge over decades with a utilitarian end in view. This reasoning is why no proponent of biodiversity is assuaged by the prospect of biotechnology being able to create new life forms or clone existing ones; nor of the growing capacity of zoos, aquariums, and botanical gardens to preserve genetic material or individual specimens ex situ for possible cloning or reintroduction into the wild (as promising and potentially valuable as all these techniques are). Similarly, no number of programmatically invented languages (Esperanto, Volapük, Klingon, etc.) will suffice for lost "natural" languages.

By the same token, it is not enough to argue for biocultural diversity exclusively on economic grounds, though the economic benefits are immense. Assuredly, we are economic actors in many, probably most, of our daily decisions. But there is little doubt that in questions of ultimacy, economics is pushed aside by emotion. Any philosophy that attempts to deny the validity of experience "by explaining away its objects or translating them into terms of no emotional pertinency leaves the mind with little to care or act for," observed James. "This is the opposite condition from that of nightmare, but when acutely brought home to consciousness it produces a kindred horror. In nightmare we have motives to act, but no power; here we have powers, but no motives" (James 1890, 2:313).[29]

Now we can finally glimpse the crisis at its deepest level: Extinction is a primal nightmare, both psychologically within ourselves and out in the world apart. Our reaction to the crisis must be founded on the realization that ours is the first generation with the capacity to reverse the age-old momentum toward fecundity in both nature and culture.[30] "Extinction shuts down the generative process," warns Rolston. "The wrong that humans are doing, or allowing to happen through carelessness, is stopping the historical flow in which the vitality of life is laid" (Rolston 1985, 723). That flow and vitality are surely both biological and cultural at the same time. Evolution is the model *par excellence* of sustainability, and therefore its product, the diverse biocultural presence, has a primary claim to being of value in any calculus of what is or isn't sustainable for the future.

The human species did not evolve in a world of drab monotony. Our brains, the consciousness function they produce, and the cultural variety expressing that function have evolved over millions of years within a lavish, enveloping environment of biological riches.[31] And, conversely, the natural environment has been widely transformed by differentiated human action. Our evolutionary history tells us that cultural diversity is intimately related to the biological diversity of the nonhuman world. Current events tell us they face the same threats. The only effective way to meet them is with a cohesive, biocultural response. Through it we would find, at last, that unity does not require uniformity. Once this insight is truly grasped, we will finally be ready to start making real progress toward that elusive ideal of a common humanity.

NOTES

CHAPTER I

1. God and the Absolute are not necessarily synonymous. In the world's major religions, as John Hick views it, they are "different though overlapping aspects of our immensely complex human potentiality for awareness of the transcendent"; in other words, two alternative ways of cognizing what he calls "the Real." In this religious diversity, the concept of God "is experienced specifically as the God of Israel, or as the Holy Trinity, or as Shiva, or as Allah, or as Vishnu. And it is in relation to yet other forms of life that the Real, apprehended through the concept of the Absolute, is experienced as Brahman, or as Nirvana, or as Being, or as Sunyata" (Hick 1989, 245).

2. This passage originally appeared in James' article "Reflex Action and Theism," published in the *Unitarian Review,* November 1881. He also used it in *The Will to Believe* (1897, 118–119) with a few minor changes in wording.

3. "If we survey the field of human history and ask what feature all great periods of revival, of expansion of the human mind, display in common, we shall find, I think, simply this: that each and all of them have said to the human being, 'The inmost nature of the reality is congenial to *powers* which you possess.' . . . Nothing could be more absurd than to hope for the definitive triumph of any philosophy which should refuse to legitimate, and to legitimate in an emphatic manner, the more powerful of our emotional and practical tendencies" (James 1890, 2:314, 2:315).

4. James felt that religious faith was synonymous with a "faith in the existence of an unseen order of some kind in which the riddles of the natural order may be found explained. In the more developed religions the natural world has always been regarded as the mere scaffolding or vestibule of a truer, more eternal world . . ." (James 1897, 51). See also James 1902, 507, 517–518.

5. Charles Derber, a professor and author of the 1999 book *Corporation Nation: How Corporations Are Taking Over the World and What We Can Do About It,* reports that most of his students are astonished on being exposed to a critique of corporate power: "Many say they cannot imagine Disney or Microsoft being too powerful. These corporations are, after all, the source of the lifestyle pleasures and magical technology that make their lives fun and their studies easier" (cited in Rowley 1999, 31).

6. Todorov (1993) provides an excellent discussion of exoticism as a form of egotism.

7. "For a thousand years Europe had been a prey to intolerant, intolerable visionaries. The common sense of the eighteenth century, its grasp of the obvious facts of human suffering, and of the obvious demands of human nature, acted on the world like a bath of moral cleansing. Voltaire must have the credit, that he hated injustice, he hated cruelty, he hated senseless repression, and he hated hocus-pocus. Furthermore, when he saw them he knew them. In these supreme virtues, he was typical of his century, on its better side" (Whitehead 1925, 86–87).

8. Lovejoy had been a colleague of James in the latter's last years, exchanging letters with him on *Pragmatism* and *A Pluralistic Universe* (see Perry 1936, 1:719, 2:480–485, and 2:595–597). Coincidentally, the lectureship under which Lovejoy produced *The Great Chain of Being* was a memorial to James. Lovejoy took due note of this and was able, with unconcealed satisfaction, to work into his magnum opus an encomium to his old friend and teacher (Lovejoy 1936, 313–314).

9. The biologist George Gaylord Simpson maintained (1961, 148) that the belief in the fixity of species was "a relatively late theological development and had not been long accepted" when Linnaeus wrote his famous line that "there are as many species as were created in the beginning." However, elsewhere (pp. 30–31) Simpson conceded that the idea of each species having an essence, an archetype, went back to the ancient Greeks. They were the source of the belief "that the archetype is in some way the reality and that organisms are merely the shadows, reflections, or imperfect and evanescent embodiments of that transcendental reality. Archetypes were considered philosophically as Platonic *ideas*, or theologically as patterns of divine creation" (p. 49). In these statements he is in agreement with Lovejoy.

10. Whether Plato would have agreed with Lovejoy's interpretation is an open question. Lovejoy was, as far as I can tell, freely drawing inferences from the *Timaeus* on which to found the principle of plenitude (the phrase is his, not Plato's). After going through it, Lovejoy sums up: "The Intellectual World was declared by Plato to be deficient without the sensible. Since a God unsupplemented by nature in all its diversity would not be 'good,' it followed that he would not be divine. And with these propositions the simile of the Cave in the *Republic* was implicitly annulled—*though Plato himself seems never to have realized this*" (Lovejoy 1936, 52–53, emphasis added). Later (p. 56), Lovejoy ascribes to Aristotle the principle of continuity, as well as "the idea of arranging (at least) all animals in a single graded *scala naturae* according to their degree of 'perfection'" (p. 57).

11. Citing Volkmann's 1884 edition of the *Enneads*, 1:247.

12. A noteworthy recent example is Hick's *Evil and the God of Love* (1966), which, aside from offering the author's own theodicy, contains a thorough historical analysis of past attempts. See also the discussion of "person-making" in Hick 1989.

13. "Multiplicity, therefore, and variety, was needful in the creation, to the end that the perfect likeness of God might be found in things according to their measure. . . . The per-

fection of the universe therefore requires not only a multitude of individuals, but also diverse kinds, and therefore diverse grades of things" (Lovejoy 1936, 76, quoting *Summa contra Gentiles*, II:45, Rickaby's translation).

14. Law, writing in 1732, intuited that "there is no manner of *chasm* or *void*, no link deficient in this great chain of beings" so that "every distinct order, every class or species of them, is as full as the nature of it would admit, or God saw proper. There are perhaps so many in each class as could exist together without some *inconvenience* or *uneasiness* to each other" (Lovejoy 1936, 185).

15. The quote is from Lucas de la Haye, *La Vie de M. Benoît de Spinoza*, as cited in Léon Brunschvicg, *Spinoza et Ses Contemporains*, p. 333.

16. Quoting Ray's *Three Physico-Theological Discourses*, 1713, 3rd ed., 149.

17. They tried to "temporalize" it by conceding that not all possible forms of being had to exist all the time; rather, their existences could come and go, spread out across all eternity (Lovejoy 1936, 244).

18. The irony is valid only insofar as one accepts the stock view of the Enlightenment as a monolithic homogenizing program that aims to snuff out artistry, spirituality, free will, and so forth (i.e., the view summarized above by Lovejoy). In his recent book *Consilience: The Unity of Knowledge* (1998), Edward O. Wilson argues that the Enlightenment ideal is not incompatible with any of these things.

19. Not necessarily near the top, either. It was often held that there were many ranks between humans and God. There are Alexander Pope's memorable lines from *An Essay on Man* of 1734: "Placed on this isthmus of a middle state / A being darkly wise, and rudely great" etc. His contemporary, Edward Young, put it more plainly: "Distinguished link in being's endless chain / Midway from nothing to the deity" (quoted in Lovejoy 1936, 190).

20. In *The Varieties of Religious Experience*, James makes the same basic point: "The fact of diverse judgments about religious phenomena is therefore entirely unescapable, whatever may be one's own desire to attain the irreversible. But apart from that fact, a more fundamental question awaits us, the question whether men's opinions ought to be expected to be absolutely uniform in this field. Ought all men to have the same religion? Ought they to approve the same fruits and follow the same leadings? Are they so like in their inner needs that, for hard and soft, for proud and humble, for strenuous and lazy, for healthy-minded and despairing, exactly the same religious incentives are required? Or are different functions in the organism of humanity allotted to different types of man, so that some may really be the better for a religion of consolation and reassurance, whilst others are better for one of terror and reproof? It might conceivably be so; and we shall, I think, more and more suspect it to be so as we go on" (James 1902, 333). Further on (p. 487), he uses the same argument to state a positive case for religious diversity.

21. Quoting *Reden über die Religion*. However, in the next sentence it is noted that "Christianity is, indeed, for Schleiermacher, the highest of the positive religions; but its superiority lies only in its freedom from exclusiveness. It does not claim 'to be universal and to rule alone

over mankind as the sole religion'" (Lovejoy 1936, 311, citing *Reden,* V). A dubious claim, to say the least. Todorov (1993) gives several examples from French thought where expressions of universal tolerance segue seamlessly into ethnocentrism or nationalism.

22. "Probably the weightiest contribution to our feeling of the rationality of the universe which the notion of the absolute brings is the assurance that however disturbed the surface may be, at bottom all is well with the cosmos—central peace abiding at the heart of endless agitation. This conception is rational in many ways, beautiful aesthetically, beautiful intellectually (could we only follow it into detail), and beautiful morally, if the enjoyment of security can be accounted moral. Practically it is less beautiful; for . . . in representing the deepest reality of the world as static and without a history, it loosens the world's hold upon our sympathies and leaves the soul of it foreign. Nevertheless it does give *peace*, and that kind of rationality is so paramountly demanded by men that to the end of time there will be absolutists, men who choose belief in a static eternal, rather than admit that the finite world of change and striving, even with a God as one of the strivers, is itself eternal" (James 1909, 113–115).

23. James, writing in 1902: "That vast literature of proofs of God's existence drawn from the order of nature, which a century ago seemed so overwhelmingly convincing, to-day does little more than gather dust in libraries, for the simple reason that our generation has ceased to believe in the kind of God it argued for" —thanks, I would say, mostly to Darwin (James 1902, 73–74; see also 91–92).

24. Quoting *Vorlesungen über dramatische Kunst und Literatur,* 1809, in Schlegel's *Collected Works,* Vol. V, 5, 15–16.

25. Lovejoy's words here actually were written apropos of Friedrich Schlegel's "famous definition of Romantic poetry: 'die romantsiche Poesie ist eine progressive Universalpoesie.' It must be universal, not in the restrictive sense . . . ."

CHAPTER 2

1. Wilson's reference to Thomas Gray's *Elegy Written in a Country Churchyard,* one of the best-known poems in English literature, is particularly appropriate in this regard. One of the poem's themes, exemplified in the famous line "Some mute inglorious Milton here may rest," is to lament the potential lost to provinciality.

2. Cf. Engel 1993, 209.

3. In the late 1960s, the biologist Raymond Dasmann specifically linked the preservation of natural and cultural diversity in his book *A Different Kind of Country.* In fact, he cast the entire book as a plea for diversity in the face of the "homogenization of the earth under the control of a single dominant human monoculture" (Dasmann 1970, viii). About the same time, the botanist Hugh H. Iltis, worried about the long-term psychological effect of "asphalt jungles" on an increasingly urbanized populace, suggested that "biotic and cultural diversity . . . may well be an important attribute to general health, a psychiatric safety valve

that is not a luxury but an indispensable human need" (Iltis 1968, 119). Now, a generation later, cultural diversity is beginning to find its way into important conservation literature. It is given a prominent (though cursory) place in two recent worldwide plans. It is one of the elements of a "world ethic of sustainability" underpinning *Caring for the Earth*, the successor to the *World Conservation Strategy* (IUCN, UNEP, and WWF 1991, 14), and is closely linked with the definition of biodiversity given in the *Global Biodiversity Strategy* (WRI, IUCN, and UNEP 1992, 3). However, the first books to systematically and theoretically link biological and cultural diversity are Posey 1999—conceived as a supplement to UNEP's *Global Biodiversity Assessment* (Heywood 1995), in which cultural diversity was virtually ignored—and Maffi 2001b.

4. The ethicist J. Ronald Engel writes, "Biological and cultural diversity depend on one another, and in many cases must be preserved together" (Engel 1993, 189). The ethnobotanist Pei Shengji says that the practices of the indigenous Dai people of southwest China demonstrate "the co-existence of biological diversity and cultural diversity and suggested that the principle of biodiversity conservation and the conservation of cultural diversity should be considered as concomitant processes and as integral factors in the conservation of biological diversity today" (Pei Shengji 1993, 118-119).

5. For a contemporary report, see Tangley 1986.

6. See Soulé and Wilcox 1980; Soulé 1986. The discipline is linked to a cluster of related ethical precepts, clearly stated by Rey C. Stendell: "Conservation biology is based upon several underlying assumptions: (1) the diversity of organisms is good; (2) the untimely extinction of populations and species is bad; (3) ecological complexity is good; (4) evolution is good; and (5) biological diversity has intrinsic values. While there may be some disagreement on the validity of these assumptions, scientists working in this field generally accept them as ethical principles for conservation biology" (2000, 109).

7. It is commonly stated to be just that: see Groombridge 1992, xiii; Heywood 1995, 5. My argument here is that it emphatically is *not*.

8. But it has not entirely escaped them. Some indigenous people see it as a calculated, repressive political ploy (see Posey 1999b, 5–6).

9. "The practical urgency of on-the-ground conservation is based on a deeper respect for life. Extinction is forever, and when danger is ultimate, absolutes are most relevant" (Rolston 1993, 41–42).

10. Whether scientists merely pretend to objectivity in the first place is of course a perennial point of dispute between them and their critics (see Ellen and Harris 1999 for a related point about Western science's claim to universality). What is indisputable about the rise of biodiversity, as both a subject of study and as an ethical movement, is that it has allowed many scientists to straddle the bounds of objectivity and at the same time allow themselves some emotional involvement in their work. As Richard Norgaard shrewdly points out: "We have . . . seen over the past two decades thousands of biologists switch from supposedly 'being objective' and letting change fall where it may to 'having an objective,' the conser-

vation of biological diversity. They have made an overt decision to come together and 'be subjective' to work towards an objective. The conservation biologists are not only doing the science of diversity but have produced new narratives about our relation to the natural world that represent some of the finest scientific writing of the century" (Norgaard 2001, 538).

11. He was acutely aware of biological diversity's beauty, however. The few rhetorical flights he allowed himself in *The Origin of Species* have almost exclusively to do with this theme.

12. In his autobiography, Wilson gives an interesting account of his "'conversion'" to biodiversity campaigning. The problem of biodiversity destruction had troubled his mind for decades, but he sublimated it (though it welled up repeatedly in a vividly disturbing dream). Then, in 1979, he read the first published estimates of tropical rainforest loss by the conservationist Norman Myers. This primed him, but he "was finally tipped into active engagement" by the example of his friend and fellow academic scientist, the botanist Peter Raven, who "made it clear that scientists in universities and other research-oriented institutions must get involved" and not leave the task to conservationists alone. "One day on impulse I crossed the line. I picked up the telephone and said, 'Peter, I want you to know that I'm joining you in this effort. I'm going to do everything in my power to help" (Wilson 1994, 354–358, quotes on p. 357). I suspect that this sequence, from rational understanding to deepening apprehension to the crossing of a moral threshold, is the line of approach for most scientists who have committed to biodiversity.

13. It is, as William James put it in *The Will to Believe*, a "live option" involving momentous choices (1897, 2–4). See also James 1907, 286–288.

14. Cf. Peterson 1999.

15. Cf. Lyons 1999 for a Native American perspective, and Burgoa 2000 (especially p. 212) for an Andean view.

16. As, for example, from Terry Erwin ("the product of organic evolution . . . the diversity of life in all its manifestations"), Jerry Franklin ("the complete array of organisms, biologically mediated processes, and organically derived structures out there on the globe"), and so on (Takacs 1996, 46–50). Interestingly, Rosen declined to give Takacs his own definition: "I'm content with the definitions that are out there" (pp. 49–50).

17. In a similar vein, the philosopher Bryan Norton says that "the value of biodiversity is the value of everything there is. It is the summed value of all the GNPs of all countries from now until the end of the world. We know that, because our very lives and our economies are dependent upon biodiversity. If biodiversity is reduced sufficiently, and we do not know the disaster point, there will no longer be any conscious beings. With them will go all value—economic and otherwise" (Norton 1988, 205). Although this last statement would certainly not go unchallenged by some of Norton's colleagues, it is difficult to deny that the questions that have preoccupied professional philosophers—questions of epistemology, metaphysics, ontology, and the like—are rendered extremely problematic if the human race is no longer around to think them. In fact, they are made meaningless, unless there is some

other truly existing "possible world" in which they continue to be considered by sentient beings. *That* idea, at least, must for the time being remain speculative. It is far less speculative to assert, as Norton does, that biodiversity is the linchpin of humankind's continued existence.

18. "The impacts of local community resource uses on biodiversity are mixed in any given situation; they could help to maintain or conserve it, enhance it, or reduce it. An analysis of the overall impact is complicated by the fact that the same resource use might have different impacts on different elements or levels of biodiversity. For instance, agricultural practices might enhance diversity at the genetic level (e.g., by developing several landraces of a particular species), reduce it at the species level (e.g., by clearing forests for making fields), but enhance it at the ecosystem level (e.g., by creating a mosaic of microecosystems)" (Kothari and Das 1999, 189). For a fine empirical study of how groups can perceive and then manage a local environment in drastically different ways, see Atran 2001.

19. Marine animal families are so used for two reasons. First, the sample of species in the fossil record is small and biased because relatively few species, and only certain kinds of species, tend to produce fossil remains. Higher taxa such as families therefore better reflect the overall history of taxonomic diversity because enumerating families helps avoid this bias. Second, long-term changes are best documented in marine species. Seas tend to accumulate sediment, whereas land erodes; in this relatively more stable medium the fossil record is more complete. Also, marine fossil species have been studied more than terrestrial ones (see the technical discussion in Heywood 1995, 202–203). It should be remarked that marine biological diversity is still immense today: 15 of the 32 major phyla known to occupy the seas are found there and there alone, and there are dramatic differences between marine and terrestrial organisms in size, shape, and means of living. By far the greatest portion of the biosphere, measured by volume, is in the oceans. Yet the immensity of the marine realm does not exempt it from human impacts, especially in coastal and near-shore habitats, which are among the most productive on Earth. "The present loss in marine biodiversity, like that, on land, is unprecedented and alarming" (Agardy 1997, 16).

20. There is a certain amount of extinction resistance built into some groups of organisms, and, as less-resistant groups die out over time, this residuum tends to form a larger proportion of the whole (see the discussion in Heywood 1995, 208). However, humans are the first species with the technological capacity to overwhelm this otherwise-resistant residuum through wholesale change of the planet's environment, including the climate.

21. In fact, there is evidence that small-scale agriculture can enhance local biodiversity; see discussion below, and also, for a brief overview, Posey 1999.

22. Cf. Pimm, Moulton, and Justice 1995.

23. "The ongoing . . . wave of human-caused disruption to biodiversity is fundamentally different from its predecessors. Today, breeding populations are disappearing from species of all evolutionary forms and sizes, from the largest trees to the tiniest soil microbes. And they are disappearing from all regions and habitats" (Cincotta and Engelman 2000, 26).

24. Population densities in hunter–gatherer societies were on the order of 0.02 to 0.2 persons per square kilometer; in early farming societies, 25 to 1,000 persons per square kilometer, or more (Heywood 1995, 722).

25. Whose biological impacts in the Americas are treated in Crosby 1986.

26. Citing Louis Harris and Associates, "Biodiversity in the Next Millennium" survey, carried out on behalf of the American Museum of Natural History, 1998.

27. As Heywood (1995, 202) points out (using a life span of 4 million years derived from marine animal fossils and then generalized to all ecosystems), such durations seem "an incredibly long time by human cultural standards" but are really "remarkably short" in comparison to the nearly 4 billion years that life has been on Earth.

28. Nonetheless, May (2000, 42) rightly insists that "today's evolutionary heritage of living species is not a negligible fraction of those ever to have graced the planet"; as I interpret it, though today's complement of species is small in percentage terms, the amount of genetic information they embody is enormous and of incalculable significance to nature and civilization.

29. Obviously, for the "graph of life" to generally rise over geological time, as it has, the overall average (or "canonical") speciation rate must have been higher than the background extinction rate. Sepkoski (1999, 260–261) discusses this relationship, which also depends on the average duration of a species.

30. A second way is to try to project future extinction rates based on recent changes in the threat status of species on the IUCN *Red List* (May, Lawton, and Stork 1995, 16–18), and a third uses assessments of extinction probability distributions as functions of time (Mace 1995).

31. In both cases, the rate accelerated after the beginning of the Industrial Revolution: Nearly three times as many became extinct since 1810 as before (see Groombridge 1992, 200).

32. Stork (1997, 60–61) presents a more refined analysis along this same line of reasoning.

33. Many of the species recorded as becoming extinct since 1600 are from islands: another built-in bias, because islands are especially attractive to researchers. It is sobering to realize that these extinctions—undercounted though they may be—are quite probably the mere tail-end of earlier, much heavier extinctions. Pimm, Moulton, and Justice (1995) have estimated that the first human colonists of the central Pacific eliminated about half of the species on each island group. If this is true elsewhere, then a tremendous number of scientifically undescribed species have become extinct at the hands of humans. Unless they left fossil remains, these extinctions will never become part of the written record.

34. See Lugo 1988 for a critique of the early estimates.

35. Because of a lack of intercollated, synoptic databases, problems with synonymy, and time lags in cataloguing, it is not possible to state with certainty the number of species that have been formally described. Noting all this, Robert M. May gives a more conservative esti-

mate of 1.5 million, and a similarly conservative estimate of 6.8 million for the total number of species extant (2000, 32–38).

36. High-end estimates of total global species numbers are based on the likely magnitude of beetle species richness in the tropics, based on the famous canopy-fogging collections of Terry Erwin in the 1980s (for a summary, see Erwin 1988). Erwin's fellow entomologist Nigel Stork is skeptical of such high estimates, and has published a careful and thorough critique (Stork 1997, 50–55), helpfully citing in full the source text for Erwin's original estimate, which originally appeared in the rather difficult-to-find specialist journal *Coleopterists's Bulletin* (Erwin 1982). See also May 2000 (37–38).

37. As has "biodiversity." In American conservative political circles it is regarded, with some justification, as a code word for a constellation of political attitudes that oppose economic development and capitalism. Some public officials, sensitive to the potential for backlash, have, with equal calculation, shied away from using the word, turning instead to more palatable terms (e.g., "the creation," which happens also to be shrewdly religious).

38. The International Forum on Globalization (http://www.ifg.org) is one current source for the anti-globalization movement.

39. Over thirty years ago, an author described how, "in the Congo, I have seen radio bulletins picked up by transistor radios and tapped out on talking drums into the swamp forests. In the dyak long-houses of Borneo, youngsters awaiting their pagan initiation into manhood listen to the Beatles. . . . On the breathless heights of the altiplano in South America, the disinherited heirs of the Incas watch Disney films," and so forth (Calder 1966). Recent critiques of globalization are full of similar, up-to-date examples (e.g., Barber 1995).

40. See Skutnabb-Kangas 2000 (425–436) for a refutation.

41. Anthropologists, ethnographers, and sociologists routinely study a multitude of variables related to social organization. They include division of labor, land tenure, kin groups and descent, kinship terminology, rules for marriage and inheritance, political organization and modes of social control, types of domestic unit, and ways of resolving conflicts (Levinson 1991, xvii). The ethnographer George Peter Murdock (1967; 1981) codified some of these variables for more than 1,250 cultures.

42. Thrupp 1999 gives other examples.

43. Cf. the idea of "genetic distance" as it relates to language distribution, discussed in many publications by Luigi Luca Cavalli-Sforza and co-workers (e.g., Cavalli-Sforza 2000).

44. Based on the first edition of the *World Christian Encyclopedia* (Barrett 1982). Unfortunately, the second (2001) edition does not provide a current (year 2000) global population summary for the 71 ethnolinguistic families, as did the first edition. It would be possible to derive current global figures from the new edition, but doing so would require adding up the population numbers that are given separately for each of the 12,583 constituent peoples that make up the 71 ethnolinguistic families (see Barrett, Kurian, and Johnson 2001, 2:26–241).

45. "This figure is based on several assumptions: first, that religion began with the Nean-

derthals, who about 100,000 years ago were carefully burying their dead with grave goods and building small altars of bear bones in caves; second, that there have been at all times since the Neanderthals a thousand or more culturally distinct human communities, each with its own religion; and third, that in any cultural tradition, religions change into ethnographically distinct entities at least every thousand years" (Wallace 1966, 3–4). This last statement is interesting in that it mirrors "the crude glottochronological formula that each language splits into two languages every 1,000 years. . ." (Mühlhäusler 1995, 155). Morris Swadesh, one of the first linguists to evince a serious interest in the causes of linguistic diversity, was also a proponent of glottochronology, which has since been largely discredited.

46. That is, all forms above the level of idiolect.

47. The classic statement of this is by Benjamin Lee Whorf (1956; original work 1941), who put the case in exceptionally compelling terms. Despite many attempts to consign this hypothesis to the scrap heap, it is still very much alive as a topic of debate.

48. The limitations of the *Ethnologue* data, which would apply to any global language survey under current conditions, are discussed in Harmon 1995.

49. Assuming the respondents understood what was meant by "mother tongue"; for real-world examples to the contrary from the 1961 census in India, see Skutnabb-Kangas 2000, 11.

50. There are additional problems with adopting language as the barometer for cultural diversity. It is very hard, in many cases impossible, to distinguish distinct languages from other language forms such as dialects (a point taken up again in Chapter 3). A more accurate view of global language richness would include dialects in the tally. The trouble is that right now we are even more in the dark about the global status of dialects than we are about distinct languages. The current *Ethnologue Index* (Grimes 2000b) lists more than 41,000 dialects and variant names for the languages reported in the *Ethnologue* itself, but does not give a separate figure for dialects alone. (The 1992 *Index* listed 37,000 dialect/variant names; Grimes 1992a.) In addition, many *Ethnologue* entries for individual languages make reference to the existence of dialects, sometimes dozens of them, without giving their names, so these must be added to the *Index* total. And *Ethnologue* does not even attempt to list all the dialects of major world languages such as English. It is safe to assume there are tens of thousands of dialects in use. In any event, dialects (like their biological counterpart, subspecies) are as intrinsically valuable as are languages (Wolfram and Schilling-Estes 1995).

51. "Just before the agricultural revolution, there were perhaps 5–10 million people" (Ehrlich 1995, 221, citing Biraben 1979).

52. Both figures come from my database analysis of the 1992 *Ethnologue* (Harmon 1995). The median was calculated by taking the 5,729 languages with reported mother-tongue speaker totals, sorting the database by those totals (lowest to highest), and finding the 2,865th language in the sort: Bubi, spoken in Gabon, with an estimated 5,000 speakers. The average was calculated by taking the cumulative total of speakers reported for the 5,729 languages and dividing: 5.369 billion speakers ÷ 5,729 languages = 937,618, rounded down to 900,000.

53. The *Ethnologue* total consists largely of oral languages; some experts on deaf languages believe that there are thousands of unenumerated sign languages in use, and suggest that the true number of living languages exceeds 10,000 (Skutnabb-Kangas 2000, 216). The Linguasphere database (published online at www.linguasphere.org, and summarized in print in Barrett, Kurian, and Johnson 2001, 2:245–529) lists about 13,500 languages, but its classification system, which has only recently been published, has yet to achieve widespread acceptance.

54. Undoubtedly additional extinctions during the period were offset by new languages coming into being, and countless diachronic language transformations (such as the change from Middle English to Modern English) have occurred.

55. In a paper given at a seminar at the University of Bristol in April 1995, Mark Pagel estimated the total number of languages ever spoken (as opposed to the average number spoken during a given period) to be somewhere between a low of 31,000 and a high of 600,000, with 140,000 as a middle figure (reported in Ostler 1995, 6). I have not been able to read the original paper, but if Pagel is correct, this would mean that between 80 percent and 99 percent of all languages that ever existed are now extinct—a proportion comparable with that of species extinctions.

56. At www.sil.org/ethnologue.

57. The increase in living languages over the figure reported in the 1992 (12th) edition (Grimes 1992b, cited below) is mainly attributable to dialects being reclassified as languages in their own right, not to new discoveries (Grimes 2000a, vii).

58. More conservative estimates run closer to 5,000 (Ruhlen 1991, 1), and a consensus figure of 6,000 has been proposed (Krauss 1992, 5); but again, see Skutnabb-Kangas on this point (2000, 216)—she argues forcefully that deaf languages (among others) are being seriously undercounted and "invisibilized." And, as noted above, the Linguasphere database runs to 13,500 languages. Until these debates are settled, for my own part I report here the numbers given in the 1992 edition of *Ethnologue*, not because they are infallible (no one, least of all SIL, says they are), nor because they are the latest figures, but simply because they are the figures I have analyzed in detail in previous papers (Harmon 1995; Harmon 1996). As a result, I incline to a working figure of 6,500. See Skutnabb-Kangas (2000, 16–46) for a thorough discussion of the problems inherent in trying to count the world's languages.

59. The following is condensed from Harmon 1995, which describes in detail the structure of the shadow database, assumptions made in calculating speaker totals, limitations of the data, and so on.

60. At the time of the study, 1995, the world's population stood at about 5.75 billion; now it is over 6 billion and of course still climbing. I do not think the increase makes any significant difference in the findings reported here, since probably the largest part of the increment has accrued to the largest languages.

61. Actually, this assumption is by no means entirely safe. Breton, as Krauss (1992, 7) has pointed out, may be drifting toward moribundity even though it had perhaps 1 million speak-

ers within living memory. This is because the number who use it every day is far lower than a million, with the number who know only Breton lower yet, and the number of children who speak it as a first language even lower still (Kuter 1989, 75–76). Even large regional languages with seemingly stable situations are subject to rapid decline under the pressure from one of the "world languages." "This process can take place within one or two generations after a long period of linguistic stability during which the native language hardly seemed threatened. Quechua, with an estimated 8.5 million speakers . . . is a case in point. After having occupied, for centuries, an unchallenged position in a vast part of western South America, accompanied by an unprecedented expansion into areas where it had not been in use before, it is now rapidly losing ground in one of its original strongholds, central Peru, as a probable result of social and economic developments affecting that region. While older people are still fluent in the language and monolinguals can be found here and there, the younger generation has no more than a passive knowledge, if any, of the language" (Adelaar 1991, 50).

62. We get an inkling of the range of qualitative distinctions—here arrayed into something close to a continuum—in this passage by Leanne Hinton, pondering the future for California Indian languages: "Must we say that Hebrew or Latin ever really 'died'? Or can we say instead that even when there were no native speakers, people studied them and used them and often learned to speak them fluently. And even, in the case of Hebrew, brought the language back into native status when the conditions were right for it. No matter what we do, it appears to be the fate of most California Indian languages to lose the last of their native speakers within the next few decades. But many of them have a good chance of remaining alive as languages of study, as second languages, even as fluently spoken second languages, and perhaps even as second languages learned in early childhood for some communities" (Hinton 1995, 42). Cf. Hymes 1971, ix.

63. Quoting Jenner 1876, 194–195.

64. The whole process from premoribundity to extinction took about 175 years. It would be interesting to see whether this chronology holds for other recently extinct languages as well. I suspect the overall course of events is now running much faster.

65. Yn Cheshaght Ghailckagh, Thie ny Gaelgey, St. Judes, Ramsey, 1M7 2EW Isle of Man.

66. Or, as Skutnabb-Kangas consistently and unflinchingly refers to it, will be *murdered* by political forces that are against multilingualism. She insists that the terminology of "language death" trivializes and falsifies the process by implying that speech communities passively give up their language in favor of a more powerful one, instead of there being an active human agency, as expressed through unjust power relationships, as the real cause. She calls this *linguicide* (see Skutnabb-Kangas 2000, passim).

67. The worldviews and subsistence practices of many indigenous peoples do not even distinguish between nature and culture (for numerous examples, see Posey 1999 and the papers thereinafter). They have thus completely internalized the reciprocity, putting their cultures even more acutely at risk if it is lost.

68. "Nature may be said to be evolved from the monotony of non-existence by the creation of diversity. It is plausibly asserted the we are conscious only so far as we experience difference" (Jevons 1877, 173).

69. For a comprehensive overview of the biocultural perspective, see Maffi 2001b.

70. Dasmann (1970, 30) and Wilson (1992, 351) have touched upon the importance of diversity as being characteristic of the milieu of human evolution.

71. "The bottom of being is left logically opaque to us, as something which we simply come upon and find, and about which (if we wish to act) we should pause and wonder as little as possible" (James 1897, 73).

CHAPTER 3

1. Though this chapter was unmistakably inspired by the *Origin*, Lyell's book as a whole left Darwin "deeply disappointed" because (as he said in a contemporaneous letter to Joseph Hooker) the author's "timidity" prevented him from expressing definitive support for evolution by natural selection (Darwin 1892, 268).

2. Citing, for example, Max Müller's *Comparative Mythology* of 1856.

3. Sound recording capable of playback was not invented until 1877.

4. Lyell also realized that, just as in natural selection, language change through cultural selection has an element of randomness in it, some linguistic innovations being made deliberately, others, "casually and as it were fortuitously" (Lyell 1863, 468).

5. Published 1871. See quotes in Ruhlen 1991, 25; Ruhlen 1994, 136–137.

6. Indeed, academic linguistics has come to be dominated by semiotic and structuralist/generative grammarian theories that either de-emphasize or simply deny any connection between language and the rest of existence. For analyses from a biocultural viewpoint, see Corbett 2001 and Pawley 2001; for a sharp critique, Wollock 2001.

7. Cf. Skutnabb-Kangas 2000, 215–219, 365–374.

8. Setting aside the issue that "choice" in regard to language use does not mean *fully free choice* in any practical sense. Language use is constrained by many social and political factors (discussed in Skutnabb-Kangas 2000).

9. Which also was Lyell's view, as I read him.

10. Lyell again: "[T]he learned are not agreed as to what constitutes a language as distinct from a dialect. Some believe that there are 4,000 living languages, others that there are 6,000, so that the mode of defining them is clearly a mere matter of opinion. Some contend, for example, that the Danish, Norwegian, and Swedish form one Scandinavian tongue, others that they constitute three different languages, others than the Danish and Norwegian are one—mere dialects of the same language, but that Swedish is distinct" (Lyell 1863, 458).

11. "Nationalism plays a big role in people's conception of language identities" is how the linguist Morris Swadesh bluntly put it. He went on to contrast the mutual intelligibility of Urdu and Hindi—which are marked as separate languages because of political, ethnic,

and religious boundaries (Urdu being associated with Pakistan, the Arabic tradition, and Islam; Hindi, with India, the Indic tradition, and Hinduism and Buddhism)—with the relatively mutual unintelligibility of the Venetian and Neapolitan "variants" of Italian. "Certain cases are of course obvious. No one would hesitate to say that Spanish and English are different languages, with little mutual intelligibility, or that Chicago and St. Louis English are variants of the same language. Between the two clear extremes, however, the scale of variation is continuous" (Swadesh 1971, 14, 15). It could be argued that in most cases the "obviousness" of a speech form's status as either language or dialect is debatable. For example, it is apparently not obvious to many people, even nearby in Scandinavia and Russia, that there are several Sámi languages, rather than dialects of a single one. Turning the tables, it might be perfectly obvious to a native speaker of one of these Sámi languages that "there is one German language, spoken across Germany, Britain, Sweden, Norway, Denmark, Iceland, the Netherlands, and so on, and it exists as several local dialects" named German, English, and so forth (Skutnabb-Kangas 2000, 197).

12. Recent benchmark overviews of species concepts and speciation are Otte and Endler 1989 and Howard and Berlocher 1998.

13. Darwin was as perplexed as anyone on this score, as a reading of his discussion of "doubtful species" in the *Origin* will disclose (Darwin 1859, 44–54). His frustration boiled over in an 1853 letter to Joseph Hooker: "After describing a set of forms, as distinct species, tearing up my M.S., and making them one species; tearing that up and making them separate, and then making them one again (which has happened to me) I have gnashed my teeth, cursed species, and asked what sin I have committed to be so punished . . ." (Burkhardt 1996, 129).

14. Coyne and Orr (1999, 1), though mindful of competing concepts, declare the biological species concept (which they abbreviate as "BSC") to be "the most useful" of the bunch, noting that "the recent burst of work on speciation reflects almost entirely the efforts of those adhering to the BSC. In fact, *every* recent study on the 'genetics of speciation' is an analysis of reproductive isolation."

15. Or downright incoherent; see Sneath and Sokal 1973, 18.

16. Another basic difference between the biological and phylogenetic species concepts is that the former assumes that "closed gene pools are entities that have for the most part been irreversibly launched on an independent course of evolution by mutation and recombination" whereas the latter "reflects rigorously the history of groups of related species without reference to their hypothesized future" (Wilson 2000, 24).

17. A "morph" is a distinguishable sympatric and synchronic interbreeding population of a single species (Simpson 1961, 178).

18. Also see Simpson 1961 and Heywood 1995 for representative discussions of the many different concepts of species put forth over the years.

19. "Cultural diversity is as essential for human cultural evolution as genetic diversity is for biological evolution and therefore for the long-term survival of human societies" (Shankar 1999, 535).

20. Dixon was inspired by the counter-Darwinian evolutionary theory of punctuated equilibria first put forth by Stephen Jay Gould and Niles Eldredge in the 1970s.

21. Cf. Norton 1987 (207), as quoted below in Chapter 5.

22. This is a strict definition, adopted for ease of data manipulation within a large database. It excludes languages that are, for all practical purposes, endemic; for example, languages for which *Ethnologue* reports a single tiny emigrant community. A more realistic definition of "endemic language" might include languages that have, say, 95 percent of their speakers in a single country. For more, see Harmon 1995, 10–11.

23. As noted above, this discussion is based on a previous paper (Harmon 1995) that analyzed the 1992 edition in detail. Because the patterns to be discussed here are so overwhelming, I am convinced that my analysis of the latest (14th) edition (Grimes 2000a, 2000b)—now in progress—will not change my main conclusions.

24. As will be discussed shortly, geography and environment are not the sole determinants of language distribution; see Nettle's case study discussion (1999a, 70–74, and the sources cited therein) of the Tsembaga Maring people for a sense of the extremely complicated interplay between social, climatic, and geographical factors in Papua New Guinea.

25. Nettle's study (1999a) is to be admired for boldly invoking ecological concepts in explaining global linguistic diversity—something that is, regrettably, still mindlessly branded as "ecological determinism" in certain quarters of the social sciences. However, based on the evidence he presents in the book, I do not think that he is justified in claiming that ecological risk is a *universal* constraint; e.g., his statement on p. 94 that "no factor has been as strong or general as ecological risk" (see pp. 69–70 and 79–96 for Nettle's general discussion of his model). For a number of methodological reasons (see pp. 82–83), Nettle excludes from his analysis all nontropical countries—as well as some key tropical or subtropical ones, such as Mexico and Brazil, that are exceedingly rich in languages. Given the exclusions, claims that his model is globally applicable must be considered premature.

26. The source for the EBA data (Stattersfield et al. 1998, 38) only gives the top 19 countries. The authors also give the 25 countries with the most restricted-range species (essentially, endemic bird species). Here, the concurrence with the top language countries is 15 of 25, or 60 percent (Stattersfield et al. 1998, 36).

27. In a similar vein, linguists have used species terminology as a tool to help reconstruct both the migrations of speech communities and the structure of proto-languages (Hinton 1994, 87–92).

28. Writing of place-based knowledge, the environmental philosopher Bryan G. Norton concludes that "if cultures become more homogenous [*sic*] we can expect landscapes to become more homogenous as a direct and indirect result. Conversely, if we hope to save the biodiversity of the earth, we must also save the special dialectics that have emerged in these local places" (Norton 1999, 471).

29. The adaptations of which I speak are sociopolitical. A language tends to become immured within the power relationships of its culture. It should be kept in mind that the

reasons why some languages are large and widespread and others are restricted to a village or two probably have little, if anything, to do with the structure of the language. For instance, one could plausibly argue that English dominates TV and the Internet because of what happened on September 13, 1759, the day the British defeated the French and their allies on Quebec's Plains of Abraham in the decisive battle of the French and Indian (Seven Years') War. This sealed English as the language of the soon-to-be United States, where, a couple of centuries down the road, TV and the Internet happen to have been developed, for reasons unrelated to any structural advantages of English itself. That is to say, the technologies could just as plausibly have developed within a French-speaking America.

30. See Adelaar 1991 (45) and Crosby 1986 (196–216) for South America, Garza Cuarón and Lastra 1991 (105) for Mexico, Zepeda and Hill 1991 (136), for the United States, Kinkade 1991 (157) for Canada, and Denevan 1992 for the Western Hemisphere in general.

31. Even if people stay in rural communities, a similar alienating effect can occur if the natural environment of the area is replaced by intensive monoculture farming or livestock production. The planners in charge of one of the largest ecological restorations ever attempted, the Guanacaste National Park project in northwestern Costa Rica, have factored this into their plans. Here, where almost all the native dry tropical forest has been replaced by farms and ranches, the planners are seeking to re-establish some of the biocultural ties to the land that were destroyed in the switchover. Without such ties, the planners believe, appreciation for the project's objectives will never arise, and the children of the area will end up being stranded "without either their predecessors' contact with the natural world nor the cultural offerings of the large cities that are supported by their parents' agriculture" (Janzen 1988b, 244).

32. See Orr 1993 (415–440) for a critique of what might be called "chic disdain for nature" among urbanophiles.

33. The abiotic contribution to biological and cultural diversity is immense. Consider, for example, geology. One way geological diversity is expressed is through geomorphology (landforms). Landforms exhibit vast differences in elevation, longitude (think of the implications of there being no comparably vast austral counterpart to the boreal regions), and soil composition, to name just three. As a kind of meta-biospheric diversity, there are the differences in global processes: climate, inorganic chemical cycling, energy flows, water balance regimes, detoxification functions, carbon sequestration, and so forth (see Daily 1997). The interaction of all these with biological and cultural diversity is what makes up landscapes, which are important ecologically and culturally. I cannot even pretend to do this topic justice here. For good, short discussions of geology's place in ecosystem management, see Hughes, Popenoe, and Geniac 1999; and Smith 1999. For the relevance of the general physical environment to biodiversity, see Handawela 2000.

34. For a discussion of Nichols's views on linguistic lineage diversity and their relevance to the human origination-migration debate, see Harmon 1996; for a recent dissent to her

views on the latter, see Nettle 1999a, 1999b. See Nettle 1999a (115 n. 1) for the use of the term "phylogenetic" instead of "genetic" level with respect to linguistic diversity.

35. This is already being done on a limited (but increasing) basis (Berlocher 1998, 10–11). See also Wilson 2000 (24).

36. "It seems to be a general human tendency to stress the sharpness of distinctions between classes and to overemphasize gaps in the spectrum of phenetic [i.e., observed] variation. Mutually exclusive classes are frequently used conceptually by humans, although we are repeatedly warned against stereotyping events and individuals. Nevertheless we succumb to a natural tendency to avoid intersecting sets, which would result in some individuals being members simultaneously of more than one set" (Sneath and Sokal 1973, 200). It seems, however, that Sneath and Sokal may be guilty of unconscious ethnocentrism here. There is much evidence that the worldviews of various indigenous peoples are far more tolerant of conceptual ambiguity; numerous examples are found in the volume *Cultural and Spiritual Values of Biodiversity* (Posey 1999a). See also Robson 1928 (25) on the blurring of properties at the margins.

37. "In spite of Aristotle's recognition of the multiplicity of possible systems of natural classification, it was he who chiefly suggested to naturalists and philosophers, of later times, the idea of arranging (at least) all animals in a single graded *scala naturae* according to their degree of 'perfection'" (Lovejoy 1936, 58). Looking to the future at the very end of the *Origin*, Darwin ventured that his theories would one day lead systematists to "weigh more carefully and to value higher the actual amount of difference" between putative species, so that science will be "freed from the vain search for the undiscovered and undiscoverable essence of the term species" (Darwin 1859, 485). See also James 1890, 2:335.

38. "Every naturalist knows vaguely what he means when he speaks of a species" (Darwin 1859, 44).

39. Darwin himself did not hesitate to admit that "the existence of groups would have been of simple signification, if one group had been exclusively fitted to inhabit the land, and another the water; one to feed on flesh, another on vegetable matter, and so on; but the case is widely different in nature; for it is notorious how commonly members of even the same sub-group have different habits" (Darwin 1859, 411). This remark falls within (what is to my mind) one of the most important sections of the *Origin*, the discussion of classification at the start of Chapter 13.

40. A point that, incidentally, testifies to the desirability of maintaining diversity: consilience can operate only when the overall field of knowledge is broad and varied.

41. Cf. the discussion of contiguity and similarity in Simpson 1961 (4).

42. "Not only can we employ the idea of likeness in this manner, but we apply it incessantly and universally to the whole mass and train of our sensations" (Whewell 1840, 1:453).

43. It is instructive to compare Wittgenstein on this point with what Darwin has to say about the complexity of family relationships (not only of species but, significantly for our

discussion, of languages) in the *Origin* (1859, 420–423; the diagram referred to therein appears at the end of the book in the Harvard 1964 facsimile of the first edition).

44. There is now a large interdisciplinary literature on boundaries and "boundary space"; see Schonewald 2001 for an overview.

45. Thousands of works have been written about him and his philosophy. For exegesis of the family resemblance concept, see Baker and Hacker 1980 (315–353), who, incidentally, recognize both Whewell and James as precursors of Wittgenstein on this point (325). Baker and Hacker also provide a good discussion of Wittgenstein's thinking on "*Merkmal*- [= 'characteristic'] definitions": that is, definitions in the essentialist tradition.

46. Beckner uses the terms "polytypic" and "monotypic"—a poor choice, because both carry a different meaning elsewhere in taxonomy. For a discussion, see Simpson (1961, 94), who did not propose an alternative. I follow Sneath and Sokal (1973, 20–21, citing earlier work by Sneath) and Lumsden and Wilson (1981, 27) in using "polythetic" and "monothetic."

47. He then goes on to acknowledge Wittgenstein's family resemblance formulation.

48. "It is no less the case in the classification of culturgens [equivalent to what Richard Dawkins calls 'memes'] than in the classification of organisms that the choice of the level of discrimination used to define the taxonomic category must be arbitrary" (Lumsden and Wilson 1981, 29). See also Darwin (1859, 47), Robson (1928, 6), Gilmour (1940, 468–469), and Sneath and Sokal (1973, 69) for definitions to the effect that a species is whatever a competent systematist says it is, and Darwin (1859, 419) on the "almost arbitrary" value of genera, order, families, and so forth. For his part, Beckner conceded that, in order to effectively define a polythetic class, one must have "at least a rough criterion of membership" beforehand (1959, 24)—circular as well as arbitrary, it would seem. I wonder, though, how arbitrary all this truly is; perhaps these classificatory decisions are instead enmeshed in extremely complex, yet rules-based, decision regimens that are encoded partly genetically and partly culturally, and whose depths we have not yet plumbed (à la Lumsden and Wilson 1981). Sneath and Sokal (1973, 146) approach that conclusion: "But when all is said and done, the validation of a similarity measure by the scientists working in a given field has so far been primarily empirical, a type of intuitive assessment of similarity based on complex phenomena of human sensory physiology."

49. Skutnabb-Kangas (2000, 155–165) explains how an individual's ethnic, linguistic, and political identities are often constructed through a kind of negotiation between how one identifies one's self ("endo-definition") and how one is identified by others ("exo-definition"), producing an "ambo-categorisation."

50. When the characteristics in common are both few and very general, then the entities in question constitute a singularity only in the most abstract sense, as James noted: "[T]wo compound things are similar when some one quality or group of qualities is shared alike by both, although as regards their other qualities they may have nothing in common. The moon is similar to a gas-jet, it is also similar to a foot-ball; but a gas-jet and a foot-ball are not simi-

lar to each other. When we affirm the similarity of two compound things, we should always say *in what respect it obtains*. Moon and gas-jet are similar in respect of luminosity, and nothing else; moon and foot-ball in respect of rotundity, and nothing else. Foot-ball and gas-jet are in no respect similar—that is, they possess no common point, no identical attribute" (James 1890, 1:579).

51. The key to understanding this paradox is recognizing the paramount agency of modification with descent: "We can understand why a species or a group of species may depart, in several of its most important characteristics, from its allies, and yet be safely classed with them. This may be safely done, and is often done, as long as a sufficient number of characters [Beckner's Proviso again!], let them be ever so unimportant, betrays the hidden bond of community of descent. Let two forms have not a single character in common, yet if these extreme forms are connected together by a chain of intermediate groups, we may at once infer their community of descent, and we put them all into the same class" (Darwin 1859, 426). In the next chapter we will return to this point, generalized to an ultimate degree, in the guise of James's view of a pluralistic, concatenated universe.

52. "The perfect demonstration of speciation is presented by the situation in which a chain of intergrading subspecies forms a loop or overlapping circle of which the terminal links have become sympatric without interbreeding, even though they are connected by a complete chain of intergrading or interbreeding populations" (Mayr 1963, 507). Mayr was referring to geographic (= allopatric) speciation, which he championed to the exclusion of other explanations.

53. So called from the title of a volume of state-of-the-art papers edited by Julian Huxley in 1942. See Wilson 1994 (112) and Berlocher 1998 (6–8) for the pantheon of geneticists and evolutionary biologists who were responsible for the Synthesis.

54. Cf. Darwin 1859, 175.

55. As it was then styled; now *Felis catus*.

56. Later on, Dobzhansky characterizes the modal points in more technical terms, building on ideas put forth by his fellow geneticist Sewall Wright. To summarize: If we consider the total number of potential gene combinations for a particular organism (which is the "'field' within which evolutionary changes can be enacted"), it is apparent that the adaptive values for some will be greater than for others. Probably a vast majority of the combinations are "discordant and unfit for survival." Closely similar gene combinations will tend to be similar in adaptive value. "If, then, the field of the possible gene combinations is graded with respect to adaptive value, we may find numerous 'adaptive peaks' separated by 'valleys.' The 'peaks' are the groups of related gene combinations that make their carriers fit for survival in a given environment; the 'valleys' are the more or less unfavorable combinations. Each living species or race may be thought of as occupying one of the available peaks in the field of gene combinations" (Dobzhansky 1941, 337). A peak, then, is a topographical image of the modal point around which a particular species clusters, while the valleys represent the discontinuities between species. As the cumulative genotypic system rep-

resented by the individuals of the species responds to mutations or selection pressures, the ensuing changes will either (1) enable the species to spread down the slope and even occupy other nearby peaks (if the variance proceeds in a nonadaptive direction, thus making the species less specialized), or (2) "force the species to withdraw to the highest level of its adaptive peak" (if the variance makes the species become more specialized). A change in the environment, if radical enough, can even, as it were, erode the peak out from under the species, calling for a complete reconstruction of the species' collective genotype if it is to avoid extinction. Species with a large amount of genetic variability in their populations will have a much better chance of adapting to the environmental change (Dobzhansky 1941, 338–339). Thus are the adaptive peaks associated with the species' modal points. The notion of adaptive peaks has become a staple of evolutionary biology (Wilson 1975, 24, 67–68).

57. The work of the 19th-century statistician Adolphe Quetelet provided the foundation for the pivotal discovery of the analysis of variance by Dobzhansky's fellow pioneer of the Modern Synthesis, the geneticist R. A. Fisher (Anastasi 1958, 6–8, 28; see also Eiseley 1958, 227).

58. "Species are not so much like lines of latitude and longitude as like mountains and rivers, phenomena objectively there to be mapped. The edges of all these natural kinds will sometimes be fuzzy, to some extent discretionary. We can expect the one species will slide into another over evolutionary time. But it does not follow from the fact that speciation is sometimes in progress that species are merely made up, instead of found as evolutionary lines articulated into diverse forms, each with its more or less distinct integrity, breeding population, gene pool, and role in its ecosystem" (Rolston 1985, 721–722).

59. Impressive though his argument is, I have a number of serious reservations about how Hick develops it, stemming from my personal views on the utility of religion as well as because of his undue emphasis on "post-axial" religious forms (i.e., the large "world religions") which promise personal salvation and liberation, leading him to give short shrift to indigenous ("pre-axial") spiritual beliefs that tend to emphasize harmony, Earth stewardship, and so forth. But the important point here is that he has plausibly characterized religious diversity.

60. "According to [Erik] Allardt, there are *no* criteria for inclusion in an ethnic group that *all* the members of the group have to fulfill. But it is necessary that *some* members fulfill *all* the criteria, and *every* member must fulfill at least *one* criterion." This is an almost perfect description of a polythetic group with a monothetic core—"a firm and stable nucleus whereas the boundaries are fluid and constantly changing" (Skutnabb-Kangas 2000, 174, paraphrasing and citing Allardt, Miemois, and Starck 1979, 12.)

CHAPTER 4

1. "Nature is a spectacle continually exhibited to our senses, in which phenomena are mingled in combinations of endless variety and novelty. Wonder fixes the mind's attention;

memory stores up a record of each distinct impression; the powers of association bring forth the record when the like is felt again. By the higher faculties of judgment and reasoning the mind compares the new with the old, recognizes essential identity, even when disguised by diverse circumstances, and expects to find again what was before experienced. It must be the ground of all reasoning and inference that *what is true of one thing will be true of its equivalent*, and that under carefully ascertained conditions *Nature repeats herself*" (Jevons 1877, 1–2). Jevons is paraphrasing *The Senses and the Intellect* (which was first published in 1855), citing Bain's three properties as central to the acquisition of knowledge.

2. "As Professor Bowen has said, 'The first necessity which is imposed upon us by the constitution of the mind itself, is to break up the infinite wealth of Nature into groups and classes of things, with reference to their resemblances and affinities, and thus to enlarge the grasp of our mental faculties, even at the expense of sacrificing the minuteness of information which can be acquired only by studying objects in detail. The first efforts in the pursuit of knowledge, then, must be directed to the business of classification. Perhaps it will be found in the sequel, that classification is not only the beginning, but the culmination and the end, of human knowledge" (Francis Bowen, *A Treatise on Logic, or, the Laws of Pure Thought*, published 1866, p. 315, as cited in Jevons 1877, 674–675). Bowen was a professor of Moral Philosophy at Harvard for many years.

3. It is not insignificant that Jamesian pragmatism depends on the ability to perceive differences: "The pragmatic rule is that the meaning of a concept may always be found, if not in some sensible particular which it directly designates, then in some particular difference in the course of human experience which its being true will make" (James 1911b, 60).

4. And Deduction versus Induction: "Rationalists are the men of principles, empiricists the men of facts; but, since principles are universals, and facts are particulars, perhaps the best way of characterizing the two tendencies is to say that rationalist thinking proceeds most willingly by going from wholes to parts, while empiricist thinking proceeds by going from parts to wholes. . . . Rationalists prefer to deduce facts from principles. Empiricists prefer to explain principles as inductions from facts" (James 1911b, 35).

5. "The difference between monism and pluralism is perhaps the most pregnant of all the differences in philosophy" (James 1897, viii), "the final question of philosophy" (James 1907, 293). Cf. Jevons 1877, 8.

6. In reference to James's thought, one can speak only metaphorically of the "substance" of consciousness because he considered consciousness to be a function, not an entity. See the chapter "Does Consciousness Exist?" in James 1912.

7. Cf. Hick 1989, 162.

8. When James wrote that the sense of sameness is the "very keel and backbone of our thinking," he not only went straight to the core truth about diversity, he did so using metaphors whose simplicity is both the source of the phrase's power and its potential downfall: The images are so conventional that we tend to pass over them without a second thought. Yet "keel" and "backbone" are rich in apt associations. The purpose of a keel is

to keep a ship tracking true as it navigates across the limitless plane of the sea. James thus deftly suggests the age-old metaphor of life as journey while sounding echoes of the questing nature of humankind, of the necessity of change, and, on a deeper level, of volition, purposefulness, and free will. The keel provides guidance and stability while enabling a freedom of movement that releases us from stasis and the moribundity that stasis implies. The "backbone" image further solidifies, and literally internalizes, the idea of the sense of sameness. The backbone holds the body up, lets us stand and walk erect like human beings. The backbone is integral to us physically and to the identity of our species. By this simple anatomical metaphor, we understand that the sense of sameness is not merely an abstract quality; it is part of our corporeal selves. So, by means of a deceptively simple utterance, James makes us feel that the ability to distinguish "same" from "other" is not just the basis of all thought, but a key to what it means to be human. As for James's declaration of the importance of the sense of sameness, it has been seconded by the contemporary philosopher W. V. Quine: "There is nothing more basic to thought and language than our sense of similarity; our sorting of things into kinds" (Quine 1969, 116).

9. Cf. Jevons 1877, 1–4; James 1912, 77; Whitehead 1925, 47.

10. See also James 1907, 129–130.

11. "Not to demand intimate relations with the universe, and not to wish them satisfactory, should be accounted signs of something wrong. Accordingly when minds of this type reach the philosophic level, and seek some unification of their vision, they find themselves compelled to correct that aboriginal appearance of things by which savages are not troubled. That sphinx-like presence, with its breasts and claws, that first bald multifariousness, is too discrepant an object for philosophic contemplation" (James 1909, 33).

12. He pointed to Spinoza, with "his barren union of all things in one substance," and Hume, "with his equally barren 'looseness and separateness' of everything," as antipodal examples of violators (James 1897, 67). He elsewhere remarked that nothing could be more dogmatically pluralistic than Hume's philosophy: "He makes events rattle against their neighbors as dryly as if they were dice in a box" (James 1911b, 198).

13. In *A Pluralistic Universe* James quoted Hegel to this effect: "'The so-called maxim of identity,' he [Hegel] wrote, 'is supposed to be accepted by the consciousness of every one. But the language which such a law demands, "a planet is a planet, magnetism is magnetism, mind is mind," deserves to be called silliness. No mind either speaks or thinks or forms conceptions in accordance with this law, and no existence of any kind whatever conforms to it. We must never view identity as abstract identity, to the exclusion of all difference. That is the touchstone for distinguishing all bad philosophy from that alone which deserves the name of philosophy. If thinking were no more than registering abstract identities, it would be a most superfluous performance. Things and concepts are identical with themselves only in so far as at the same time they involve distinction'" (James 1909, 94, citing *Smaller Logic*, 184–185). See also *The Principles of Psychology*, where James says that no two things are "in

scientific rigor identical. We call those of them identical whose difference is unperceived. Over and above this we have a *conception* of absolute sameness, it is true, but this, like so many of our conceptions . . . is an ideal construction got by following a certain direction of serial increase to its maximum supposable extreme" (James 1890, 1:533).

14. To James, whether the samenesses we arrive at actually and independently exist or not is less important than the common acceptance of our intentions regarding them. "Without the psychological sense of identity, sameness might rain down upon us from the outer world for ever and we be none the wiser. With the psychological sense, on the other hand, the outer world might be an unbroken flux, and yet we should perceive a repeated experience. Even now, the world may be a place in which the same thing never did and never will come twice. The thing we mean to point at may change from top to bottom and we be ignorant of the fact. But in our meaning itself we are not deceived; our intention is to think of the same. The name which I have given to the principle, in calling it the law of constancy in our meanings, accentuates its subjective character, and justifies us in laying it down as the most important of all the features of our mental structure" (James 1890, 1:460).

15. For example, the quotes given in the main text above, as well as James 1912, 65; James 1890, 1:402–403.

16. "This world *might* be a world in which all things differed, and in which what properties there were were ultimate and had no farther predicates. In such a world there would be as many kinds as there were separate things. We could never subsume a new thing under an old kind; or if we could, no consequences would follow. Or, again, this might be a world in which innumerable things were of a kind, but in which no concrete thing remained of the same kind long, but all objects were in a flux. Here again, though we could subsume and infer, our logic would be of no practical use to us, for the subjects of our propositions would have changed whilst we were talking. In such worlds, logical relations would obtain, and be known (doubtless) as they are now, but they would form a merely theoretic scheme and be of no use for the conduct of life. But our world is no such world. It is a very peculiar world, and plays right into logic's hands. *Some* of the things, at least, which it contains are of the same kind as other things; *some* of them remain always of the kind of which they once were; and some of the properties of them cohere indissolubly and are always found together. *Which* things these latter things are we learn by experience in the strict sense of the word, and the results of experience are embedded in 'empirical propositions.' Whenever such a thing is met with by us now, our sagacity notes it to be of a certain kind; our learning immediately recalls that kind's kind, and then *that* kind's kind, and so on; so that a moment's thinking may make us aware that the thing is of a kind so remote that we could never have directly perceived the connection. The flight to this last kind *over the heads of the intermediaries* is the essential feature of the intellectual operation here." (James 1890, 2:651–652; see also James 1907, 139–140)

I would say that the opposing theoretical worlds James describes at the start of this quote *are* indeed chaos, and just so because of the reasons he gives; whereas the real world he contrasts them with, and which he thinks is primordially chaotic, is *not*.

17. James holds it to be an "undeniable fact" that *"any number of impressions, from any number of sensory sources, falling simultaneously on a mind* WHICH HAS NOT YET EXPERIENCED THEM SEPARATELY, *will fuse into a single undivided object for that mind*. The law is that all things fuse that *can* fuse, and nothing separates except what must" (James 1890, 1:488). This I would say holds true subconsciously.

18. True chaos ensues only when we are unable to process raw diversity into the usual background feeling of rightness, whether because of mental illness, extraordinary emotional stress, and the like.

19. As a literary technique, known as "stream of consciousness."

20. Conception cuts up the perceptual flux in a fundamentally necessary yet purely ideal way (James 1911b, 48–50).

21. "The substitution of concepts and their connections, of a whole conceptual order, in short, for the immediate perceptual flow, thus widens enormously our mental panorama. Had we no concepts we should live simply 'getting' each successive moment of experience, as the sessile sea-anemone on its rock receives whatever nourishment the wash of the waves may bring. With concepts we go in quest of the absent, meet the remote, actively turn this way or that, bend our experience, and make it tell us whither it is bound. We change its order, run it backwards, bring far bits together and separate near bits, jump about over its surface instead of plowing through its continuity, string its items on as many ideal diagrams as our mind can frame. All these are ways of *handling* the perceptual flux and *meeting* distant parts of it; and as far as this primary function of conception goes, we can only conclude it to be what I began by calling it, a faculty superadded to our barely perceptual consciousness for its use in practically adapting us to a larger environment than that of which brutes take account. We *harness* perceptual reality in concepts in order to drive it better to our ends." (James 1911b, 64–65)

22. "Trains of concepts unmixed with percepts grow frequent in the adult mind; and parts of these conceptual trains arrest our attention just as parts of the perceptual flow did, giving rise to concepts of a higher order of consciousness" (James 1911b, 51). Nonetheless, our transition to adulthood almost invariably means we lose a characteristically childlike ability to wonder, and with it goes access to certain realms of existence that we can never fully regain entry to, try as we would. "What was bright and exciting becomes weary, flat, and unprofitable. The bird's song is tedious, the breeze is mournful, the sky is sad. . . . From one year to another we see things in new lights. What was unreal has grown real, and what was exciting is insipid. The friends we used to care the world for are shrunken to shadows; the women, once so divine, the stars, the woods, and the waters, how now so dull and common; the young girls that brought an aura of infinity, at present hardly distinguishable existences;

the pictures so empty; and as for the books, what *was* there to find so mysteriously significant in Goethe, or in John Mill so full of weight? Instead of all this, more zestful than ever
is the work, the work; and fuller and deeper the import of common duties and of common goods" (James 1890, 1:232, 1:233–234; see pp. 2:400–402 for more vivid descriptions
in this vein). To grasp the "fulness" of existence, we need, James felt, both perception and
conception: "Percepts and concepts interpenetrate and melt together, impregnate and fer-
tilize each other. Neither, taken alone, knows reality in its completeness. We need them
both, as we need our legs to walk with" (James 1911b, 53).

23. "But whether about generalities or particulars, man thinks always by the same methods. He observes, discriminates, generalizes, classifies, looks for causes, traces analogies, and
makes hypotheses" (James 1911b, 15). All these, of course, are acts of comparison.

24. Linguists will recognize echoes of this in the work of Benjamin Lee Whorf.

25. My guess is that, as the neurological sciences discover more and more about the structure and functioning of the brain, we will eventually be able to explain parts of the process
in more precise physical terms. I don't think James would disagree: About halfway through
the first volume of *The Principles of Psychology*, after he has strode through whole realms of
the discipline, he casually lets this statement fall: "I trust that the student will now feel that
the way to a deeper understanding of the order of our ideas lies in the direction of cere-
bral physiology" (James 1890, 1:387).

26. See Perry (1936, 2:112–132, 2:173) for an account of James's turn away from psychology. Perry emphasizes that James did not entirely drop psychology after the publication
of the *Principles* (as is sometimes said), although it is true that he never produced another
original work on the subject.

27. Perry described James's approach to knowledge this way: "The mind is encouraged
to imagine and speculate freely, its merit like that of Darwin's organic world, being its fecundity. But such profligacy can be tolerated only because in the end experience, wielding
the razor of Occam, is to prune away every irrelevance and superfluity. . . . The wider the
range of alternatives from which experience may choose, the richer the truth which experience will yield" (Perry 1936, 1:563). The concentration of the powers of the mind on that
which it deems relevant and essential—which is to say, the wielding of the mental razor,
or, as I have called it, the distillation of sameness—must start from a realm of abundance
situated in the imagination. The goal of the imagination, from its seat in diversity, is to race
free. Taking Perry one step further, I would say that the freely speculating imagination is
not just analogous to the fecundity of Darwin's organic world; it is *informed* by it. From
this comes the biocultural presence. And, as Perry says, the wider the range of diversity, the
greater the depth of insight that our choices may gain for us (and the fact that we can choose
is important). However, exposure to diversity doesn't guarantee anything. Here again
diversity carries the baggage of contingency, and there is no promise of an inevitable
outcome.

28. "But the whole feeling of reality, the whole sting and excitement of our voluntary life, depends on our sense that in it things are *really being decided* from one moment to another, and that it is not the dull rattling off of a chain that was forged innumerable ages ago" (James 1890, 1:453).

29. James did not so hesitate, declaring that "creatures extremely low in the intellectual scale may have conception. All that is required is that they should recognize the same experience again. A polyp would be a conceptual thinker if a feeling of 'Hollo! thingumbob [*sic*] again!' ever flitted through its mind" (James 1890, 1:463).

30. See also Quine 1969, in which he repeatedly declares that a sense of comparison is an endowment all animals share. This is the sense meant by Simpson when he speaks of association by contiguity as opposed to association by similarity (1961, 4), and James makes the point explicitly: "Compared with men, it is probable that brutes [i.e., nonhuman animals] neither attend to abstract characters, nor have association by similarity. Their thoughts probably pass from one concrete object to its habitual concrete successor far more uniformly than is the case with us. In other words, their associations of ideas are almost exclusively by contiguity. . . . [T]he most elementary single difference between the human mind and that of brutes lies in this deficiency on the brute's part to associate ideas by similarity—characters, the abstraction of which depends on this sort of association, must in the brute always remain drowned, swamped in the total phenomenon which they help constitute, and never used to reason from" (James 1890, 2:348, 2:360; see also 2:353).

31. Although James came to regret titling *The Will to Believe* as he did (he later said he should have called it *The Right to Believe*; see Perry 1936, 2:244–245) the use of the word "will" does emphasize that belief is a matter of choice, that some impulse drives us to believe, and that we ought to defend the right to refer our beliefs to this capacity.

32. Elsewhere he was more sober: "Intellectualism has its source in the faculty which gives us our chief superiority to the brutes, our power, namely, of translating the crude flux of experience of our merely feeling-experience into a conceptual order. . . . Both theoretically and practically this power of framing abstract concepts is one of the sublimest of our human prerogatives. We come back into the concrete from our journey into these abstractions, with an increase both of vision and of power. It is no wonder that earlier thinkers, forgetting that concepts are only man-made extracts from the temporal flux, should have ended by treating them as a superior type of being, bright, changeless, true, divine, and utterly opposed in nature to the turbid, restless lower world. The latter then appears as but their corruption and falsification" (James 1909, 217–218).

33. "Things tell a story. Their parts hang together so as to work out a climax. They play into each other's hands expressively. Retrospectively, we can see that altho [*sic*] no definite purpose presided over a chain of events, yet the events fell into a dramatic form, with a start, a middle, and a finish. In point of fact all stories end; and here again the point of view of a many is the more natural one to take. The world is full of partial stories that run parallel to one another, beginning and ending at odd times. They mutually interlace and interfere at

points, but we can not unify them completely in our minds. . . . It follows that whoever says that the whole world tells one story utters another of those monistic dogmas that a man believes at his risk" (James 1907, 143–144).

34. See also James 1907, 145–146.

35. The citation is of Santayana's review in *The Atlantic Monthly* (65), 553 (1891). James himself expressed it this way: "And when our idealists recite their arguments for the Absolute, saying that the slightest union admitted anywhere carries logically absolute One-ness with it, and that the slightest separation admitted anywhere logically carries disunion remediless and complete, I cannot help suspecting that the palpable weak places in the intellectual reasonings they use are protected from their own criticism by a mystical feeling that, logic or no logic, absolute Oneness must somehow at any cost be true." Thus, "the world's oneness has generally been affirmed abstractly only, and as if any one who questioned it must be an idiot" (James 1907, 154–155, 159).

36. James buttressed his pluralism by developing a doctrine he called "radical empiri-cism"—an empiricism that recognizes that the conjunctive relations between things are every bit as real as the things themselves, and as real as the disjunctions between them. See, in general, his *Essays in Radical Empiricism* (James 1912); also, for example, *The Will to Believe* (James 1897, ix), *A Pluralistic Universe* (James 1909, 79), and *The Meaning of Truth* (James 1911a, xvii).

37. And whole ecosystems, in fact. Darwin provided an excellent illustration of con-catenation in nature with his description of how the number of house cats determines the abundance of such flowers as heartsease in a particular area through the intermediation of mice and bumblebees (Darwin 1859, 73–74).

38. See James 1890 (2:648) for a discussion of the mechanism, which he called the "prin-ciple of mediate subsumption," by which this works.

39. "The essence of life is its continuously changing character; but our concepts"—on which absolutists over-rely—"are all discontinuous and fixed, and the only mode of making them coincide with life is by arbitrarily supposing positions of arrest therein" (James 1909, 253). Such over-conceptualizing robs life of its vivacity: ". . . to understand life by concepts is to arrest its movement, cutting it up into bits as if with scissors, and immobilizing these in our logical herbarium where, comparing them as dried specimens, we can ascertain which of them statically includes or excludes which other. This treatment supposes life to have already accomplished itself, for the concepts, being so many views after the fact, are retro-spective and post mortem" (James 1909, 244).

40. "Once admit that the passing and evanescent are as real parts of the stream [of thought] as the distinct and comparatively abiding; once allow that fringes and halos, inar-ticulate perceptions, whereof the objects are as yet unnamed, mere nascencies of cognition, premonitions, awarenesses of direction, are thoughts *sui generis*, as much as articulate imag-inings and propositions are; once restore, I say, the *vague* to its psychological rights, and the matter presents no further difficulty" (James 1890, 1:478–479). See also James 1890, 1:255–256; and James 1911a, 35).

41. This statement predates Einstein, of course, who published his landmark works on the theory of relativity, including the space–time continuum, in 1918 and 1921.

42. "'The world is One,' therefore, just so far as we experience it to be concatenated. One by as many definite conjunctions as appear. But then also *not* One by just as many definite *dis*junctions as we find. The oneness and the manyness of it thus obtain in respects which can be separately named. It is neither a universe pure and simple nor a multiverse pure and simple" (James 1907, 148).

43. "Contrast is the reproductive phase of the first law of mind—*relativity*, or Discrimination. Everything known to us is known in connexion with something else, the opposite or negation of itself: light implies darkness; heat supposes cold. Knowledge, like consciousness, in the last resort, is a transition from one state to another; and both states are included in the act of knowing either. . . . The 'great' is great only because there is something else 'not great,' or 'small': even when we imagine we are looking at the single property greatness, we have in our minds by *implication* the alternative, smallness; and it is only like reversing the magnet, to pass to the *explicit* consideration of the alternative—in which case, the other, 'greatness,' would be the implied property. This is what we do, when we pass from one member of a contrast to the other: both members must be in consciousness or within easy reach of consciousness, although we make only one the explicit object of consideration for the time. That the other member is still before us in a manner, is shown by the fact that, if we have been long absent from the express consideration of the alternative, we become oblivious to the force of the principal. The effect of summer warmth continued for a length of time, is to diminish the sense of warmth; a few wintry days interpolated would revive the poignancy of the sensation. *When a meaning is but dimly perceived by any one, the fault most frequently lies in the non-recognition of the opposite—that is, the thing to be excluded or denied,—the supplying of which renders the notion luminous at once.*" (Bain 1904, 599–600, emphasis added)

James declared that aesthetic and practical interests guide us in dissociating "the elements of originally vague totals": "Now, a creature which has few instinctive impulses, or interests, practical or aesthetic, will dissociate few characters and will, at best, have limited reasoning powers; whilst one whose interests are very varied will reason much better." These diverse interests lead on to a diversification of experiences, whose accumulation becomes a condition for some of the most common modes of abstract reasoning (James 1890, 2:344–345; see also 1:506–507).

44. At various places in the *Origin*, Darwin makes a related point: the tendency, if left unchecked, of dominant species to drive out weaker ones, resulting in a more homogeneous system. "We can so far take a prophetic glance into futurity as to foretel [*sic*] that it will be the common and widely-spread species, belonging to the larger and dominant groups, which will ultimately prevail and procreate new and dominant species" (Darwin 1859, 489; see also 75–76).

45. Cf. James 1912, 89–90. This is the basic idea behind Whewellian consilience.

46. Cf. Teilhard, Le Roy, and Vernadsky's idea of the "noösphere" ("mind-sphere," described by Teilhard as "the psychically reflexive human surface" of the planet) as counterpart to the atmosphere, biosphere, lithosphere, etc. (Teilhard de Chardin 1956, 103).

47. Skutnabb-Kangas (2000) speaks of the "consciousness industry."

48. "Although, when you have a continuum given, you can make cuts and dots [ellipses] in it *ad libitum*, enumerating the cuts and dots will not give you your continuum back" (James 1911b, 84).

49. Quoting Ribot's *Les Maladies de la Mémoire* (1881, 46).

50. See also James 1890, 2:316; James 1897, 70, 220–221; James 1912, 63–64.

51. Cf. Simpson 1949.

CHAPTER 5

1. "Thus we have a rather intricate system of necessary and immutable *ideal truths of comparison*, a system applicable to terms *experienced* in any order of sequence or frequency, or even to terms never experienced or to be experienced, such as the mind's imaginary constructions would be. These truths of comparison result in *Classifications*. It is, for some unknown reason, a great aesthetic delight for the mind to break the order of experience, and class its materials in serial orders, proceeding from step to step of difference, and to contemplate untiringly the crossings and inosculations of the series among themselves" (James 1890, 2:646–647).

2. The classic statement, C. P. Snow's celebrated lecture on the "two cultures," did much to bring the split before the public, but, as he said, the rift was evident (in England at least) by the 1930s and its beginnings certainly go much further back (Snow 1959, 19). Actually, Snow's main concern in the lecture was not to contrast science with the humanities in general, but the attitude of scientists with that of the literary set in particular. The image of two opposing cultures was a conscious simplification on his part (Snow 1959, 9–10).

3. Some are willing to try. In his examination of the human sociobiology debate, the philosopher of science Michael Ruse not only sees opportunities for a rapprochement between anthropologists and evolutionary biologists, he feels that a compromise position would be salutary for both disciplines (Ruse 1979, 161–162, 170–182).

4. Cf. "transformative value" as a rationale for preserving wild species (Norton 1987, chap. 10, esp. 210–211).

5. The unspoken assumption being that such efforts are objective and universally valuable. Cf. Norgaard 2001, 538. James was clear on where the boundaries of science fall: "*Moral questions* immediately present themselves as questions whose solution cannot wait for sensible proof. A moral question is a question not of what sensibly exists, but of what is good,

or would be good if it did exist. Science can tell us what exists; but to compare the *worths*, both of what exists and of what does not exist, we must consult not science, but what Pascal calls our heart. Science herself consults her heart when she lays it down that the infinite ascertainment of fact and correction of false belief are the supreme goods for man. Challenge the statement and science can only repeat it oracularly, or else prove it by showing that such ascertainment and correction bring man all sorts of other goods which man's heart in turn declares. The question of having moral beliefs at all or not having them is decided by our will. Are our moral preferences true or false, or are they only odd biological phenomena, making things good or bad for *us*, but in themselves indifferent? How can your pure intellect decide? If your heart does not *want* a world of moral reality, your head will assuredly never make you believe in one." (James 1897, 22–23; see also James 1890, 2:639)

6. Bryan G. Norton describes a similar pitfall: "What is the value of biodiversity? The question is deceiving in its simplicity, and it wise to avoid assuming, prior to careful analysis, the form that its answer will take. Note first that the definite pronoun, 'the,' already suggests there is a single answer—that however many ways people use, enjoy, worship and respect wild living things, there must be a unitary or at least synoptic answer to the question of biodiversity values. This suggestion of unitary value has gained credibility in many circles from a fear of relativism—the view that every valuation of wild life is so conditioned by local situations that judgments of nature's value add up to nothing more than the subjective and irreconcilable feelings of many different persons in many different cultures" (Norton 1999, 468).

7. Cf. Lovejoy's assessment of Romanticism: "The [Romantic] revolt against standardization of life easily becomes a revolt against the whole conception of standards. The God whose attribute of reasonableness was expressed in the principle of plenitude was not selective; he gave reality to all the essences. But there is in man a reason which demands selection, preference, and negation, in conduct and in art. To say 'Yes' to everything and everybody is manifestly to have no character at all. The delicate and difficult art of life is to find, in each new turn of experience, the *via media* between two extremes: to be catholic *without* being characterless; to have and apply standards, and yet to be on guard against their desensitizing and stupefying influence, the tendency to blind us to the diversities of concrete situations and to previously unrecognized values; to know when to tolerate, when to embrace, and when to fight. And in that art, since no fixed and comprehensive rule can be laid down for it, we shall doubtless never attain perfection. All this has now, no doubt, become a truism; but it is also a paradox, since it demands a synthesis of opposites. And to Schiller and some of the Romanticists its paradoxical aspect made it seem not less but more evidently true." (Lovejoy 1936, 312)

8. "We find life handed on, through ills and all, by wisdom genetically programmed, as well as in the cultural heritage of our forebears. The secret of life is only penultimately in the DNA, the secret of life is this struggling on, this struggling through to something higher" (Rolston 1993, 55).

9. Whether any species concept really can be applied to such organisms as viruses is a matter of contention among biologists—see, however, Wilson 2000 (23), for a specific definition of bacterial species based on DNA analysis—but that is beside the point here.

10. The difference is rooted in our being the sole moral agents on the planet. The evolved complexity of our consciousness endows us with special and unique powers over (and responsibilities to) other species, whereas the claims of justice are reciprocal with our fellow human beings.

11. We have discussed this in the context of James's philosophy, Dobzhansky's model of organic diversity, and Hick's "pluralistic hypothesis" of religion, but similar insights are found further afield. There is Whitehead's dictum on quantum mechanics: "For when we penetrate to these final entities, this startling discontinuity of spatial existence discloses itself. . . . Accordingly, in asking where the primordial element is, we must settle on its average position at the centre of each period" (1925, 53, 54). And Tzvetan Todorov, discussing Rousseau's *First and Second Discourses*, says that "Rousseau repeatedly advocates the paradoxical enterprise of discovering [universal] properties by way of difference" (1993, 11).

12. Darwin sensed something like this in his premonitions of Mendelian genetics: "Hence it seems that, on the one hand, slight changes in the conditions of life benefit all organic beings, and on the other hand, that slight crosses, that is crosses between the males and females of the same species which have varied and become slightly different, give vigour and fertility to the offspring. But we have seen that greater changes, or changes of a particular nature, often render organic beings in some degree sterile; and that greater crosses, that is crosses between males and females which have become widely or specifically different, produce hybrids which are generally sterile in some degree. I cannot persuade myself that this parallelism is an accident or an illusion. Both series of facts seem to be connected together by some common but unknown bond, which is essentially related to the principle of life" (Darwin 1859, 267).

13. "Is cultural diversity a value for its own sake and, if so, why? Cultures confer upon people their fundamental identity, their meaning, their worth, and their sense of place in the overall cosmic order. Therefore, the active defence of cultural diversity with its varied meaning systems and symbolic beliefs is essential to human development. Cultural diversity is a value for its own sake because free human persons and human communities are values in themselves. Human persons do not live except within cultural communities. Hence if a unitary paradigm of life in community is to be imposed from the requirements of a particular view of technical efficiency, that reductionist model is highly destructive of true [i.e., ethical] development" (Goulet 1993, 31).

14. See Goulet 1993 (37–38) for a discussion of biological and cultural diversity as prerequisites for ethical development and as values in their own right; also Fishman 1989 (15); Wilson 1984.

15. Whether one *values* the process is another story. According to Hick, Zen Buddhism teaches that the human propensity for "continually distinguishing, comparing and evalu-

ating" obscures the true nature of reality by falsely placing the individual consciousness at the center of existence. The process thus becomes a "distorting screen" through which the world is viewed, and only by "ending or suspending this self-centred discriminative activity" can we finally experience the world as it truly is (Hick 1989, 288–289). In contrast, I think the continual discriminative process is something to be understood and affirmed, not overcome and negated, but in any event I construe the Zen viewpoint as confirming the *humanness* of the process itself.

16. I hope that by now the reader will clearly see that such a statement is no mere rhetorical flourish, for what it means to "be human" is not simply a matter of biology, but of sociocultural conditions as well. Indeed, the definition of "humanness" could soon be revamped, depending on the collective sociocultural response to the unprecedented choices in biotechnology that are in the offing. Now that the human genome has been basically decoded, it may well soon be possible to do "germ line genetic engineering": the alteration of the hereditarily transmitted germ line to manipulate and select for specific characteristics in eggs, sperm, or very early embryos. The consequences of doing so would be momentous, and some proponents of the technology speak openly and glowingly of the coming "post-human" world (Hayes 2001, 28–30).

17. "Any discussion of the genetics of speciation must begin with the observation that species are real entities in nature, not subjective human divisions of what is really a continuum among organisms. . . . The strongest evidence for the reality of species is the existence of distinct groups living in sympatry (separated by genetic and phenotypic gaps) that are recognized consistently by independent observers" (Coyne and Orr 1999, 1).

18. "Organic diversity is an observational fact more or less familiar to everyone. It is perceived by us as something apart from ourselves, independent of the working of our minds" (Dobzhansky 1941, 3). See also the discussions of deconstruction in Soulé and Lease (1995, passim) and of subjectivism in Whitehead (1925, 130).

19. "We are trying to save the knowledge that the forests and this planet are alive—to give it back to you who have lost the understanding" (Bepkororoti Paiakan, Kayapo chief, Brazil, cited in Posey 1999b, 16). See also Norton 1987, 167–168, and esp. p. 212).

20. See the discussion of the evolutionary implications of linguistic diversity in Maffi 1999 and Maffi and Skutnabb-Kangas 1999.

21. "Cultural diversity is as essential for human cultural evolution as genetic diversity is for biological evolution and therefore for the long-term survival of human societies" (Shankar 1999, 535).

22. For a serious criticism of Allen's joke, see Orr 1993.

23. To judge from her essay "Against Nature," the writer Joyce Carol Oates is just about a perfect example of the type. For her, nature is something people possess: "Most of the time it's just an activity, a sort of hobby, a weekend, a few days, perhaps a few hours, staring out the window at the mind-dazzling autumn foliage of, say, Northern Michigan, being rendered speechless—temporarily—at the sight of Mt. Shasta, the Grand Canyon, Ansel Adams's

West. . . . Nature as the self's (flattering) mirror, but not ever, no never, Nature-in-itself" (Oates 1987, 239–240).

24. ". . . the adaptive capacity of the human organism is directly a function of its biological diversity (itself deriving from neurological complexity). Diversity is a general requirement in all living things for flexible adaptation and survival in adverse conditions. [René] Dubos states that the growing trend toward mass urban settlements poses a severe threat to the capacity of human organisms to survive collectively, if and when their urban support systems are destroyed or damaged. On instrumental grounds he argues that the maintenance of diverse capabilities, which avoid being atrophied by being used in a diversity of environments involving diverse kinds of relationships with nature, is essential to human survival" (Goulet 1993, 31).

25. More specifically: "Knowledge of cultures other than our own appears, then, along with historical research, as one of the two major modalities of comparison. As such, in turn, it is not simply one method among others but the only path leading to the requisite detachment from self and to accurate knowledge of social phenomena, whatever their nature may be" (Todorov 1993, 89). Obviously there is a third major modality Todorov is ignoring here: comparison with nature. Kellert hits the mark when he observes that nonhuman species "provide us with a profound otherness for developing our knowledge of humanity, self, and society" (Kellert 1996, 32). The solution, I am suggesting, is to combine Todorov's insights with Kellert's.

26. In no small measure, my arguments in this book could be considered an extension of Wilson's biophilia hypothesis, in which he argues that humans are innately attracted to life and lifelike processes (Wilson 1984; Kellert and Wilson 1993; Kellert 1997; for testimonials from supporters, see Takacs 1996, 219–225), but which, in practice, has been interpreted as referring exclusively to nature. I believe that most people have an innate attraction to *diversity*, and that they need access to both biological and cultural diversity to fully satisfy this need. See also Norton 1987, 228–229.

27. James did not think God could prevent evil, but he did believe God to be the only antidote to a bleak scientific materialism: "You all know the picture of the last state of the universe, which evolutionary science foresees. I can not state it better than in Mr. Balfour's words [from *The Foundations of Belief*, by Arthur James Balfour]: 'The energies of our system will decay, the glory of the sun will be dimmed, and the earth, tideless and inert, will no longer tolerate the race that has for a moment disturbed its solitude. Man will go down into the pit, and all his thoughts will perish. The uneasy consciousness which in this obscure corner has for a brief space broken the contented silence of the universe, will be at rest. Matter will know itself no longer. "Imperishable monuments" and "immortal deeds," death itself, and love stronger than death, will be as if they had not been. Nor will anything that is, be better or worse for all that the labor, genius, devotion, and suffering of man have striven through countless ages to effect'. . . . This utter final wreck and tragedy is the essence of scientific materialism as at present understood. The lower and not the higher forces are

the eternal forces, or the last surviving forces within the only cycle of evolution which we can definitely see. . . . The notion of God, on the other hand, however inferior it may be in clearness to those mathematical notions so current in mechanical philosophy, has at least this practical superiority over them, that it guarantees an ideal order that shall be permanently preserved. A world with a God in it to say the last word, may indeed burn up or freeze, but we then think of him as still mindful of the old ideals and sure to bring them elsewhere to fruition; so that, where he is, tragedy is only provisional and partial, and shipwreck and dissolution not the absolutely final things" (James 1907, 103–104, 105, 106–107).

28. "The pragmatism or pluralism which I defend has to fall back on a certain ultimate hardihood, a certain willingness to live without assurances or guarantees" (James 1911a, 229).

29. James repeated this passage in *The Will to Believe* (1897, 83).

30. "Earth is a fertile planet, and in that sense, fertility is the deepest category of all, one classically reached by the category of creation" (Rolston 1993, 59; see also pp. 45–46). "Aside from the broad tendency for the expansion of life, which is also inconstant, there is no sense in which it can be said that evolution *is* progress. . . . In this sense extinction is not merely the end but also the very antithesis of progress. . . ." (Simpson 1949, 263, 244).

31. "There is probably no one in the world who is indifferent to diversity in their environment. Practically any human endeavor that we characterize as enjoyable either directly depends on, or is greatly improved by, the factor of diversity. . . . This can not be an accidental phenomenon. It is too intimately bound up with the whole human aesthetic sense and has too much influence over our behavioral reactions to be less than directly related to our very survival instincts. A logical presumption would be that in the distant past when so many attributes were being imprinted into the human animal by the uncompromising forces of natural selection, a positive reaction to natural diversity conferred a survival advantage. No doubt the same sorts of diverse landscapes which still exert a strong attraction to us now, then connoted important facts of richness, abundance, and reliability. It was not fortuitous that we grew to love these things, and it seems especially poignant that our inherent love for diversity strongly persist right to the present, when diversity is so imminently threatened with destruction" (The Nature Conservancy 1975, 33–34).

REFERENCES

Adelaar, Willem F. H. 1991. The endangered languages problem: South America. In *Endangered Languages*. Robert H. Robins and Eugenius Uhlenbeck, eds. Oxford and New York: Berg, 45–91.

Agardy, Tundi Spring. 1997. *Marine Protected Areas and Ocean Conservation.* Austin, Tex.: R. G. Landes/Academic Press.

Allardt, Erik, Karl Johan Miemois, and Christian Starck. 1979. *Multiple and Varying Criteria for Membership in a Linguistic Minority: The Case of the Swedish Speaking Minority in Metropolitan Helsinki.* Research Reports no. 21. Helsinki: University of Helsinki Research Group for Comparative Sociology.

Anastasi, Anne. 1958. *Differential Psychology: Individual and Group Differences in Behavior.* 3rd ed. New York: Macmillan.

Atran, Scott. 2001. The vanishing landscape of the Petén Maya lowlands: People, plants, animals, places, words, and spirits. In *On Biocultural Diversity: Linking Language, Knowledge, and the Environment.* Luisa Maffi, ed. Washington, D.C.: Smithsonian Institution Press, 157–174.

Bain, Alexander. 1904. *The Senses and the Intellect.* 4th ed. New York: D. Appleton & Co. (1st ed. published 1855).

Baker, G. P., and P .M. S. Hacker. 1980. *Wittgenstein: Understanding and Meaning. An analytical commentary on the Philosophical Investigations.* Chicago: University of Chicago Press.

Barber, Benjamin R. 1995. *Jihad vs. McWorld.* New York: Times Books.

Barrett, David B., ed. 1982. *World Christian Encyclopedia: A Comparative Study of Churches and Religions in the Modern World, AD 1900–2000.* Nairobi, Kenya: Oxford University Press.

Barrett, David B., George T. Kurian, and Todd M. Johnson. 2001. *World Christian Encyclopedia: A Comparative Study of Churches and Religions in the Modern World.* 2nd ed. Oxford: Oxford University Press.

Barth, Fredrik. 1969. Introduction. In *Ethnic Groups and Boundaries: The Social Organization of Culture Difference.* Fredrik Barth, ed. Boston: Little, Brown and Co., 9–38.

Beckner, Morton. 1959. *The Biological Way of Thought,* New York: Columbia University Press.

Berg, Peter. 1981. Devolving beyond global monoculture. *CoEvolution Quarterly* 32, 24–30.

Berlocher, Stewart H. 1998. Origins: A brief history of research on speciation. In *Endless*

Forms: Species and Speciation. Daniel J. Howard and Stewart H. Berlocher, eds. New York: Oxford University Press, 3–15.

Besterman, Theodore. 1969. *Voltaire*. New York: Harcourt, Brace & World.

Biraben, Jean-Noël. 1979. Essai sur l'évolution du nombre des hommes. *Population* 34(1), 13–25.

Bloomfield, Leonard. 1933. *Language*. New York: Henry Holt.

Bodley, John H. 1982. *Victims of Progress*. 2nd ed. Palo Alto, Calif.: Mayfield.

Bodley, John H. 1994. *Cultural Anthropology: Tribes, States, and the Global System*. Mountain View, Calif.: Mayfield.

Boyd-Barrett, J. O. 1982. Cultural dependency and the mass media. In *Culture, Society, and the Media*. Michael Gurevitch, Tony Bennett, James Curran, and Janet Woollacott, eds. New York: Methuen, 174–195.

Brenzinger, Matthias, Bernd Heine, and Gabrielle Sommer. 1991. Language death in Africa. In *Endangered Languages*. Robert H. Robins and Eugenius Uhlenbeck, eds. Oxford and New York: Berg, 19–44.

Burgoa, Freddy Delgado. 2000. Local knowledge and agro-ecology in Bolivia: Dialogue for the conservation of cultural and biological diversity in the Bolivian university system. In *Links Between Cultures and Biodiversity: Proceedings of the Cultures and Biodiversity Congress 2000*. Xu Jianchu, ed. Kunming, Yunnan, China: Yunnan Science and Technology Press, 208–217.

Burkhardt, Frederick, ed. 1996. *Charles Darwin's Letters: A Selection 1825–1859*. Cambridge, U.K.: Cambridge University Press.

Calder, Ritchie. 1966. Freedom begins with breakfast. *Freedom from Hunger* 7(45), 8–11.

Cavalli-Sforza, Luigi Luca. 2000. *Genes, Peoples and Languages*. Mark Seielstad, trans. New York: North Point Press.

Cincotta, Richard P., and Robert Engelman. 2000. *Nature's Place: Human Population and the Future of Biological Diversity*. Washington, D.C.: Population Action International.

Cohen, Joel E. 1995. *How Many People Can the Earth Support?* New York: W. W. Norton.

Collinge, N. E. 1990. Language as it evolves: Tracing its forms and families. In *An Encyclopaedia of Linguistics*. N. E. Collinge, ed. London and New York: Routledge, 876–916.

Corbett, Greville C. 2001. Why linguists need languages. In *On Biocultural Diversity: Linking Language, Knowledge, and the Environment*. Luisa Maffi, ed. Washington, D.C.: Smithsonian Institution Press, 82–94.

Corliss, John O. 2000. Biodiversity, classification, and numbers of species of protists. In *Nature and Human Society: The Quest for a Sustainable World*. Peter H. Raven, ed. Washington, D.C.: National Academy Press, 130–155.

Coyne, Jerry A., and H. Allen Orr. 1999. The evolutionary genetics of speciation. In *Evolution of Biological Diversity*. Anne E. Magurran and Robert M. May, eds. Oxford: Oxford University Press, 1–36.

Cracraft, Joel. 1989. Speciation and its ontology: The empirical consequences of alternative species concepts for understanding patterns and processes of differentiation. In *Speciation and its Consequences*. Daniel Otte and John A. Endler, eds. Sunderland, Mass.: Sinauer, 28–59.

Crosby, Alfred W. 1986. *Ecological Imperialism: The Biological Expansion of Europe, 900–1900*. Cambridge, U.K.: Cambridge University Press.

Daily, Gretchen C., ed. 1997. *Nature's Services: Societal Dependence on Natural Ecosystems*. Washington, D.C., and Covelo, Calif.: Island Press.

Darwin, Charles. 1859. *On the Origin of Species by Means of Natural Selection, or the Preservation of Favoured Races in the Struggle for Life*. 1st ed. London: John Murray. Facsimile reprint. Cambridge, Mass.: Harvard University Press, 1964.

Darwin, Francis, ed. 1892. *Charles Darwin: His Life Told in an Autobiographical Chapter and in a Selected Series of His Published Letters*. New York: D. Appleton & Co. Facsimile reprint under the title *The Autobiography of Charles Darwin and Selected Letters*. New York: Dover, 1958.

Dasmann, Raymond F. 1970. *A Different Kind of Country*. New York: Collier.

Denevan, William M. 1992. The pristine myth: The landscape of the Americas in 1492. *Annals of the Association of American Geographers* 82(3), 369–385.

Dimmendaal, Gerrit J. 1989. On language death in eastern Africa. In *Investigating Obsolescence: Studies in Language Contraction and Death*. Nancy D. Dorian, ed. Cambridge, U.K.: Cambridge University Press, 13–31.

Dixon, R. M. W. 1991. The endangered languages of Australia, Indonesia, and Oceania. In *Endangered Languages*. Robert H. Robins and Eugenius Uhlenbeck, eds. Oxford and New York: Berg, 229–255.

Dixon, R. M. W. 1997. *The Rise and Fall of Languages*. Cambridge, U.K.: Cambridge University Press.

Dobzhansky, Theodosius. 1941. *Genetics and the Origin of Species*. 2nd ed. Columbia Biological Series no. 11. New York: Columbia University Press.

Dorian, Nancy C. 1998. Western language ideologies and small-language prospects. In *Endangered Languages: Language Loss and Community Response*. Lenore A. Grenoble and Lindsay J. Whaley, eds. Cambridge, U.K.: Cambridge University Press, 3–21.

Durham, William H. 1990. Advances in evolutionary culture theory. *Annual Review of Anthropology* 19, 187–210.

Durrell, Martin. 1990. Language as geography. In *An Encyclopaedia of Linguistics*. N. E. Collinge, ed. London and New York: Routledge, 917–955.

Ehrlich, Paul R. 1994. Energy use and biodiversity loss. *Philosophical Transactions of the Royal Society of London B* 344, 99–104.

Ehrlich, Paul R. 1995. The scale of the human enterprise and biodiversity loss. In *Extinction Rates*. John H. Lawton and Robert M. May, eds. Oxford: Oxford University Press, 214–226.

Ehrlich, Paul R., and Anne Ehrlich. 1981. *Extinction: The Causes and Consequences of the Disappearance of Species*. New York: Random House.

Eiseley, Loren. 1958. *Darwin's Century: Evolution and the Men Who Discovered It*. Garden City, N.Y.: Doubleday.

Eldredge, Niles. 1995. Mass extinction and human responsibility. In *Biology, Ethics, and the Origins of Life*. Holmes Rolston, III, ed. Boston: Jones and Bartlett, 63–87.

Ellen, Roy, and Holly Harris. 1999. Embeddedness of indigenous environmental knowledge. In *Cultural and Spiritual Values of Biodiversity: A Complementary Contribution to the Global Biodiversity Assessment*. Darrell Addison Posey, ed. London: Intermediate Technologies and United Nations Environment Programme, 180–184.

Encyclopædia Britannica. 1990. *1990 Britannica World Data*. Chicago: Encyclopædia Britannica.

Engel, J. Ronald. 1993. The role of ethics, culture, and religion in conserving biodiversity: A blueprint for research and action. In *Ethics, Religion, and Biodiversity: Relations Between Conservation and Cultural Values*. Lawrence S. Hamilton, ed. Cambridge, U.K.: The White Horse Press, 183–214.

Erwin, Terry L. 1982. Tropical forests: Their richness in Coleoptera and other arthropod species. *Coleopterist's Bulletin 36*, 74–75.

Erwin, Terry L. 1988. The tropical forest canopy: The heart of biotic diversity. In *Biodiversity*. E. O. Wilson, ed. Washington, D.C.: National Academy Press, 123–129.

Featherstone, Mike. 1991. Global culture: Introduction. In *Global Culture: Nationalism, Globalization, and Modernity*. Mike Featherstone, ed. London: SAGE, 1–14.

Fishman, Joshua. 1982. Whorfianism of the third kind: Ethnolinguistic diversity as a worldwide societal asset (The Whorfian Hypothesis: Varieties of validation, confirmation, and disconfirmation). *Language in Society* 11, 1–14.

Fishman, Joshua. 1989. *Language and Ethnicity in Minority Sociolinguistic Perspective*. Clevedon, U.K., and Philadelphia: Multilingual Matters.

Fleischacker, Samuel. 1994. *The Ethics of Culture*. Ithaca, N. Y., and London: Cornell University Press.

Futuyma, Douglas J. 1989. Macroevolutionary consequences of speciation: Inferences from phytophagous insects. In *Speciation and its Consequences*. Daniel Otte and John A. Endler, eds. Sunderland, Mass.: Sinauer, 557–578.

Garza Cuarón, Beatriz, and Yolanda Lastra. 1991. Endangered languages in Mexico. In *Endangered Languages*. Robert H. Robins and Eugenius Uhlenbeck, eds. Oxford and New York: Berg, 93–134.

Gellner, Ernest. 1983. *Nations and Nationalism*. Ithaca, N.Y.: Cornell University Press.

Gilmour, J. S. L. 1940. Taxonomy and philosophy. In *The New Systematics*. Julian Huxley, ed. Oxford: Oxford University Press, 461–474.

Goulet, Denis. 1993. Biological diversity and ethical development. In *Ethics, Religion, and Biodiversity: Relations Between Conservation and Cultural Values*. Lawrence S. Hamilton, ed. Cambridge, U.K.: The White Horse Press, 17–39.

Gray, Andrew. 1999. Indigenous peoples, their environments and territories: Introduction. In *Cultural and Spiritual Values of Biodiversity: A Complementary Contribution to the Global*

Biodiversity Assessment. Darrell Addison Posey, ed. London: Intermediate Technologies and United Nations Environment Programme, 61–66.

Grillo, Ralph D. 1989. *Dominant Languages: Language and Hierarchy in Britain and France.* Cambridge, U.K.: Cambridge University Press.

Grimes, Barbara F., ed. 1992a. *Ethnologue Index.* Dallas: Summer Institute of Linguistics.

Grimes, Barbara F., ed. 1992b. *Ethnologue: Languages of the World.* 12th ed. Dallas: Summer Institute of Linguistics.

Grimes, Barbara F., ed. 2000a. *Ethnologue. Volume 1: Languages of the World.* 14th ed. Dallas: SIL International.

Grimes, Barbara F., ed. 2000b. *Ethnologue. Volume 2: Maps and Indexes.* 14th ed. Dallas: SIL International.

Grimes, Joseph E. 1995. Language endangerment in the Pacific. *Oceanic Linguistics* 34(1), 1–12.

Groombridge, Brian, ed. 1992. *Global Biodiversity: Status of the Earth's Living Resources.* London: Chapman & Hall. Compiled by the World Conservation Monitoring Centre.

Hale, Ken. 1992. On endangered languages and the safeguarding of diversity. *Language* 68, 1–3.

Hamelink, Cees. 1990. Information imbalance: Core and periphery. In *Questioning the Media: A Critical Introduction.* John Downing, Ali Mohammadi, and Annabelle Sreberny-Mohammadi, eds. Newbury Park, Calif.: SAGE, 217–228.

Hamp, Eric P. 1989. On signs of health and death. In *Investigating Obsolescence: Studies in Language Contraction and Death.* Nancy D. Dorian, ed. Cambridge, U.K.: Cambridge University Press, 197–210.

Handawela, J. 2000. Relevance of physical environment as a basis for biodiversity and cultural diversity: A development concept. In *Links Between Cultures and Biodiversity: Proceedings of the Cultures and Biodiversity Congress 2000.* Xu Jianchu, ed. Kunming, Yunnan, China: Yunnan Science and Technology Press, 243–249.

Harmon, David. 1995. The status of the world's languages as reported in Ethnologue. *Southwest Journal of Linguistics* 14(1/2), 1–28.

Harmon, David. 1996. Losing species, losing languages: Connections between linguistic and biological diversity. *Southwest Journal of Linguistics* 15(1/2), 89–108.

Harmon, David. 1998a. The other extinction crisis: Declining cultural diversity and its implications for protected area management. In *Linking Protected Areas with Working Landscapes Conserving Biodiversity: Proceedings of the Third International Conference on Science and Management of Protected Areas.* Neil W. P. Munro and J. H. Martin Willison, eds. Wolfville, Nova Scotia: Science and Management of Protected Areas Association, 352–359.

Harmon, David. 1998b. Sameness and silence: Language extinctions and the dawning of a biocultural approach to diversity. *Global Biodiversity* 8(3), 2–10.

Harmon, David, and Steven R. Brechin. 1994. The future of protected areas in a crowded world. *The George Wright Forum* 11(3), 97–116.

Harrison, Richard G. 1998. Linking evolutionary pattern and process: The relevance of species concepts for the study of speciation. In *Endless Forms: Species and Speciation*. Daniel J. Howard and Stewart H. Berlocher, eds. New York: Oxford University Press, 19–31.

Hayes, Richard. 2001. The quiet campaign for genetically engineered humans. *Earth Island Journal* 16(1), 28–30.

Heywood, V. H., ed. 1995. *Global Biodiversity Assessment*. Cambridge, U.K.: Cambridge University Press.

Hick, John. 1966. *Evil and the God of Love*. New York: Harper & Row.

Hick, John. 1989. *An Interpretation of Religion: Human Responses to the Transcendent*. Basingstoke, Hamps.: Macmillan.

Hill, Jane H. 2001. Dimensions of attrition in language death. In *On Biocultural Diversity: Linking Language, Knowledge, and the Environment*. Luisa Maffi, ed. Washington, D.C.: Smithsonian Institution Press, 175–189.

Hinton, Leanne. 1994. *Flutes of Fire: Essays on California Indian Languages*. Berkeley: Heyday.

Hinton, Leanne. 1995. Current issues affecting language loss and language survival in California. *Southwest Journal of Linguistics* 14(1/2), 29–42.

Hoenigswald, Henry M. 1989. Language obsolescence and language history: Matters of linearity, leveling, loss, and the like. In *Investigating Obsolescence: Studies in Language Contraction and Death*. Nancy D. Dorian, ed. Cambridge, U.K.: Cambridge University Press, 347–354.

Howard, Daniel J., and Stewart H. Berlocher, eds. 1998. *Endless Forms: Species and Speciation*. New York: Oxford University Press.

Hughes, Stuart P., James H. Popenoe, and Judy Geniac. 1999. Missing link in ecosystem management: The role of geology. In *On the Frontiers of Conservation: Proceedings of the 10th Conference on Research and Resource Management in Parks and on Public Lands*. David Harmon, ed. Hancock, Mich.: The George Wright Society, 122–125.

Hunn, Eugene S. 2001. Prospects for the persistence of 'endemic' cultural systems of traditional environmental knowledge: A Zapotec example. In *On Biocultural Diversity: Linking Language, Knowledge, and the Environment*. Luisa Maffi, ed. Washington, D.C.: Smithsonian Institution Press, 118–132.

Hymes, Dell. 1971. Introduction. *The Origin and Diversification of Language* (by Morris Swadesh). Joel Sherzer, ed. Chicago: Aldine Atherton.

Iltis, Hugh H. 1968. The optimum human environment and its relation to modern agricultural preoccupations. *The Biologist* 50(3/4), 114–125.

IUCN, UNEP, and WWF. 1991. *Caring for the Earth: A Strategy for Sustainable Living*. Gland, Switzerland: IUCN, United Nations Environment Programme, and World Wildlife Fund.

James, William. 1890. *The Principles of Psychology*. 2 vols. New York: Henry Holt & Co. Facsimile reprint. New York: Dover, 1950.

James, William. 1897. *The Will to Believe and Other Essays in Popular Philosophy*. New York: Longmans, Green & Co. Facsimile reprint. New York: Dover, 1956.

James, William. 1902. *The Varieties of Religious Experience: A Study in Human Nature, Being the Gifford Lectures on Natural Religion Delivered at Edinburgh in 1901–1902*. New York: Longmans, Green & Co.

James, William. 1907. *Pragmatism: A New Name for Some Old Ways of Thinking*. New York: Longmans, Green & Co.

James, William. 1909. *A Pluralistic Universe: Hibbert Lectures at Manchester College on the Present Situation in Philosophy*. New York: Longmans, Green & Co. Facsimile reprint. Lincoln and London: University of Nebraska Press, 1996.

James, William. 1911a. *The Meaning of Truth*. New York: Longmans, Green & Co. Facsimile reprint. Amherst, N.Y.: Prometheus, 1977.

James, William. 1911b. *Some Problems of Philosophy: A Beginning of an Introduction to Philosophy*. New York: Longmans, Green & Co. Facsimile reprint. Lincoln and London: University of Nebraska Press, 1996.

James, William. 1912. *Essays in Radical Empiricism*. New York: Longmans, Green & Co. Facsimile reprint. Lincoln and London: University of Nebraska Press.

Janzen, Daniel H. 1988a. Tropical dry forests: The most endangered major tropical ecosystem. In *Biodiversity*. E. O. Wilson, ed. Washington, D.C.: National Academy Press, 130–137.

Janzen, Daniel H. 1988b. Tropical ecological and biocultural restoration. *Science* 239, 243–244.

Jenner, Henry. 1876. The Manx language: Its grammar, literature, and present state. *Transactions of the Philological Society 1875–76*, 172–197.

Jespersen, Otto. 1922. *Language: Its Nature, Development, and Origin*. New York: Henry Holt.

Jevons, W. Stanley. 1877. *The Principles of Science: A Treatise of Logic and Scientific Method*. Reprinted with corrections, 1913. London: Macmillan.

Jowett, Benjamin, ed. 1952. *The Dialogues of Plato*. Chicago: Encyclopædia Britannica.

Kellert, Stephen R. 1996. *The Value of Life: Biological Diversity and Human Society*. Washington, D.C.: Island Press.

Kellert, Stephen R. 1997. *Kinship to Mastery: Biophilia in Human Evolution and Development*. Washington, D.C., and Covelo, Calif.: Island Press.

Kellert, Stephen R., and Edward O. Wilson, eds. 1993. *The Biophilia Hypothesis*. Washington, D.C.: Island Press.

Kellogg, Erin L., ed. 1995. *The Rain Forests of Home: An Atlas of People and Place; Part 1—Natural Forests and Native Languages of the Coastal Temperate Rain Forest*. Portland, Ore., and Washington, D.C.: Ecotrust, Pacific GIS, and Conservation International.

Kinkade, M. Dale. 1991. The decline of native languages in Canada. In *Endangered Languages*. Robert H. Robins and Eugenius Uhlenbeck, eds. Oxford and New York: Berg, 157–176.

Kothari, Ashish, and Priya Das. 1999. Local community knowledge and practices in India. In *Cultural and Spiritual Values of Biodiversity: A Complementary Contribution to the Global Biodiversity Assessment*. Darrell Addison Posey, ed. London: Intermediate Technologies and United Nations Environment Programme, 185–192.

Krauss, Michael. 1992. The world's languages in crisis. *Language* 68(1), 4–10.

Krauss, Michael. 1995a. Disappearing languages. *Society for the Study of the Indigenous Languages of the Americas Newsletter* 14(1), 5.

Krauss, Michael. 1995b. Language loss in Alaska, the United States, and the world. *Frame of Reference* 6(1), 3–5.

Kuter, Lois. 1989. Breton vs. French: Language and the opposition of political, economic, social, and cultural values. In *Investigating Obsolescence: Studies in Language Contraction and Death*. Nancy D. Dorian, ed. Cambridge, U.K.: Cambridge University Press, 75–79.

Law, Vivien. 1990. Language and its students. In *An Encyclopaedia of Linguistics*. N. E. Collinge, ed. London and New York: Routledge, 784–842.

Levinson, David. 1991. Introduction. *Encyclopedia of World Cultures*. David Levinson, ed. Boston: G. K. Hall & Co.

Lovejoy, Arthur O. 1936. *The Great Chain of Being: A Study of the History of an Idea*. Cambridge, Mass.: Harvard University Press.

Lovejoy, Thomas E. 1980. A projection of species extinctions. In *The Global 2000 Report to the President: Entering the Twenty-First Century*. G. O. Barney, ed. Washington, D.C.: Council on Environmental Quality, U.S. Government Printing Office, 328–331.

Lovejoy, Thomas E. 1981. Prepared statement. In *Tropical Deforestation: An Overview, the Role of International Organizations, the Role of Multinational Corporations*. Hearings Before the Subcommittee on International Organizations of the Committee on Foreign Affairs, U.S. House of Representatives, 96th Congress, 2nd session. Washington, D.C.: U.S. Government Printing Office, 175–180.

Lugo, Ariel E. 1988. Estimating reductions in the diversity of tropical forest species. In *Biodiversity*. E.O. Wilson, ed. Washington, D.C.: National Academy Press, 58–70.

Lumsden, Charles J., and Edward O. Wilson. 1981. *Genes, Mind, and Culture: The Coevolutionary Process*. Cambridge, Mass., and London: Harvard University Press.

Lyell, Charles. 1863. *The Geological Evidences of the Antiquity of Man; with Remarks on Theories of the Origin of Species by Variation*. 2nd ed. London: John Murray.

Lyons, Oren. 1999. All my relations: Perspectives from indigenous peoples. In *Cultural and Spiritual Values of Biodiversity: A Complementary Contribution to the Global Biodiversity Assessment*. Darrell Addison Posey, ed. London: Intermediate Technologies and United Nations Environment Programme, 450–452.

MacArthur, Robert H. 1972. *Geographical Ecology: Patterns in the Distribution of Species*. New York: Harper & Row.

MacArthur, Robert H., and Edward O. Wilson. 1967. *The Theory of Island Biogeography*. Princeton, N.J.: Princeton University Press.

Mace, Georgina M. 1994. Classifying threatened species: Means and ends. *Philosophical Transactions of the Royal Society of London B* 344, 91–97.

Mace, Georgina M. 1995. Classification of threatened species and its role in conservation planning. In *Extinction Rates*. John H. Lawton and Robert M. May, eds. Oxford: Oxford University Press, 197–213.

Maffi, Luisa. 1998. Language: A resource for nature. *Nature & Resources* 34(4), 12–21.

Maffi, Luisa. 1999. Linguistic diversity: Introduction. In *Cultural and Spiritual Values of Biodiversity: A Complementary Contribution to the Global Biodiversity Assessment*. Darrell Addison Posey, ed. London: Intermediate Technologies and United Nations Environment Programme, 21–35.

Maffi, Luisa. 2001a. Introduction: On the interdependence of biological and cultural diversity. In *On Biocultural Diversity: Linking Language, Knowledge, and the Environment*. Luisa Maffi, ed. Washington, D.C.: Smithsonian Institution Press, 1–50.

Maffi, Luisa, ed. 2001b. *On Biocultural Diversity: Linking Language, Knowledge, and the Environment*. Washington, D.C.: Smithsonian Institution Press.

Maffi, Luisa, and Tove Skutnabb-Kangas. 1999. Language maintenance and revitalization. In *Cultural and Spiritual Values of Biodiversity: A Complementary Contribution to the Global Biodiversity Assessment*. Darrell Addison Posey, ed. London: Intermediate Technologies and United Nations Environment Programme, 37–44.

Magurran, Anne E. 1988. *Ecological Diversity and its Measurement*. London: Croon Helm.

Mander, Jerry, and Edward Goldsmith, eds. 1996. *The Case Against the Global Economy and for a Turn Toward the Local*. San Francisco: Sierra Club Books.

Marty, Martin E. 1997. *The One and the Many: America's Struggle for the Common Good*. Cambridge, Mass.: Harvard University Press.

Mattelart, Armand. 1982. *Multinational Corporations and the Control of Culture: The Ideological Apparatuses of Imperialism*. Atlantic Highlands, N.J.: Humanities Press.

Mawdsley, N. M., and N. E. Stork. 1995. Species extinctions in insects: Ecological and biogeographical considerations. In *Insects in a Changing World*. R. Harrington and N. E. Stork, eds. London: Academic Press, 322–369.

May, Robert M., John H. Lawton, and Nigel E. Stork. 1995. Assessing extinction rates. In *Extinction Rates*. John H. Lawton and Robert M. May, eds. Oxford: Oxford University Press, 1–24.

May, Robert M. 2000. The dimensions of life on Earth. In *Nature and Human Society: The Quest for a Sustainable World*. Peter H. Raven, ed. Washington, D.C.: National Academy Press, 30–45.

Mayr, Ernst. 1963. *Animal Species and Evolution*. Cambridge, Mass.: The Belknap Press of Harvard University Press.

McNeely, Jeffrey A. 1995. The interaction between biological diversity and cultural diversity. Paper presented at "International Conference on Indigenous Peoples, Environment, and Development," Zurich.

Mead, Aroha Te Pareake. 1999. Sacred balance. In *Cultural and Spiritual Values of Biodiversity: A Complementary Contribution to the Global Biodiversity Assessment*. Darrell Addison Posey, ed. London: Intermediate Technologies and United Nations Environment Programme, 112–114.

Melton, J. Gordon. 1989. *The Encyclopedia of American Religions.* 3rd ed. Detroit: Gale.

Mishler, Brent D. 2001. Biodiversity and the loss of lineages. In *On Biocultural Diversity: Linking Language, Knowledge, and the Environment.* Luisa Maffi, ed. Washington, D.C.: Smithsonian Institution Press, 71–81.

Mittermeier, Russell A., Cristina Goettsch Mittermeier, and Patricio Robles Gil. 1997. *Megadiversity: Earth's Biologically Wealthiest Nations.* Mexico City: CEMEX.

Mühlhäusler, Peter. 1995. The interdependence of linguistic and biological diversity. In *The Politics of Multiculturalism in the Asia/Pacific.* David Myers, ed. Darwin, Australia: Northern Territory University Press, 154–161.

Murdock, George Peter. 1967. *Ethnographic Atlas.* Pittsburgh: University of Pittsburgh Press.

Murdock, George Peter. 1981. *Atlas of World Cultures.* Pittsburgh: University of Pittsburgh Press.

Myers, Norman. 1979. *The Sinking Ark: A New Look at the Problem of Disappearing Species.* New York: Pergamon.

Myers, Norman. 1982. Forest refuges and conservation in Africa with some appraisal of survival prospects for tropical moist forests throughout the biome. In *Biological Diversification in the Tropics.* Ghillean T. Prance, ed. New York: Columbia University Press, 658–672.

Myers, Norman. 1983. Conservation of rain forests for scientific research, for wildlife conservation, and for recreation and tourism. In *Tropical Rain Forest Ecosystems: Structure and Function.* F. B. Golley, ed. Amsterdam: Elsevier, 325–334.

Myers, Norman. 1988. Threatened biotas: 'Hotspots' in tropical forests. *Environmentalist* 8, 1–20.

Nabhan, Gary Paul, and Sara St. Antoine. 1993. The loss of floral and faunal story: The extinction of experience. In *The Biophilia Hypothesis.* Stephen R. Kellert, and Edward O. Wilson, eds. Washington, D.C.: Island Press, 229–250.

Nash, James A. 1999. On bioresponsibility. In *Cultural and Spiritual Values of Biodiversity: A Complementary Contribution to the Global Biodiversity Assessment.* Darrell Addison Posey, ed. London: Intermediate Technologies and United Nations Environment Programme, 471–474.

National Research Council. 1980. *Research Priorities in Tropical Biology.* Washington, D.C.: National Academy Press.

Nettle, Daniel. 1999a. *Linguistic Diversity.* Oxford: Oxford University Press.

Nettle, Daniel. 1999b. Linguistic diversity of the Americas can be reconciled with a recent colonization. *Proceedings of the National Academy of Sciences USA (Anthropology)* 96, 3325–3329.

Nichols, Johanna. 1992. *Linguistic Diversity in Space and Time.* Chicago: University of Chicago Press.

Norgaard, Richard B. 1988. The rise of the global exchange economy and the loss of biological diversity. In *Biodiversity*. E. O. Wilson, ed. Washington, D.C.: National Academy Press, 206–211.

Norgaard, Richard B. 2001. Possibilities after progress. In *On Biocultural Diversity: Linking Language, Knowledge, and the Environment*. Luisa Maffi, ed. Washington, D.C.: Smithsonian Institution Press, 533–539.

Norris, Mary Jane. 1998. Canada's Aboriginal languages. *Canadian Social Trends* (Winter), 8–16.

Norton, Bryan G., ed. 1986. *The Preservation of Species*. Princeton, N.J.: Princeton University Press.

Norton, Bryan G. 1987. *Why Preserve Natural Variety?* Studies in Moral, Political, and Legal Philosophy. Princeton, N.J.: Princeton University Press.

Norton, Bryan G. 1988. Commodity, amenity, and morality: The limits of quantification in valuing biodiversity. In *Biodiversity*. E. O. Wilson, ed. Washington, D.C.: National Academy Press, 200–205.

Norton, Bryan G. 1999. Nature and culture in the valuation of biodiversity. In *Cultural and Spiritual Values of Biodiversity: A Complementary Contribution to the Global Biodiversity Assessment*. Darrell Addison Posey, ed. London: Intermediate Technologies and United Nations Environment Programme, 468–471.

Noss, Reed F., and Allen Y. Cooperrider. 1994. *Saving Nature's Legacy: Protecting and Restoring Biodiversity*. Washington, D.C., and Covelo, Calif.: Island Press.

Oates, Joyce Carol. 1987. Against nature. In *On Nature: Nature, Landscape, and Natural History*. Daniel Halpern, ed. San Francisco: North Point Press, 236–243.

Oldfield, Margery L. 1984. *The Value of Conserving Genetic Resources*. Washington, D.C.: National Park Service.

Orr, David W. 1993. Love it or lose it: The coming biophilia revolution. In *The Biophilia Hypothesis*. Stephen R. Kellert and Edward O. Wilson, eds. Washington, D.C.: Island Press, 415–440.

Ostler, Nicholas D. M. 1995. Recent meetings: Bristol—Conservation of Endangered Languages. *Iatiku: Newsletter of the Foundation for Endangered Languages* 1(1), 6.

Otte, Daniel, and John A. Endler, eds. 1989. *Speciation and its Consequences*. Sunderland, Mass.: Sinauer.

Pawley, Andrew. 2001. Some problems of describing linguistic and ecological knowledge. In *On Biocultural Diversity: Linking Language, Knowledge, and the Environment*. Luisa Maffi, ed. Washington, D.C.: Smithsonian Institution Press, 228–247.

Pedersen, Holger. 1962. *The Discovery of Language: Linguistic Science in the 19th Century*. Bloomington: University of Indiana Press. Original English edition 1931.

Pei Shengji. 1993. Managing for biological diversity conservation in the temple yards and holy hills: The traditional practices of the Xishuangbanna Dai community, southwest China. In *Ethics, Religion, and Biodiversity: Relations Between Conservation and Cultural Values*. Lawrence S. Hamilton, ed. Cambridge, U.K.: The White Horse Press, 118–132.

Perry, Ralph Barton. 1936. *The Thought and Character of William James*. 2 vols. Boston: Little, Brown & Co.

Peterson, Richard B. 1999. Central African voices on the human–environment relationship. In *Cultural and Spiritual Values of Biodiversity: A Complementary Contribution to the Global Biodiversity Assessment*. Darrell Addison Posey, ed. London: Intermediate Technologies and United Nations Environment Programme, 95–98.

Pimm, Stuart L., and Thomas M. Brooks. 2000. The sixth extinction: How large, where, and when? In *Nature and Human Society: The Quest for a Sustainable World*. Peter H. Raven, ed. Washington, D.C.: National Academy Press, 46–62.

Pimm, Stuart L., Michael P. Moulton, and Lenora J. Justice. 1995. Bird extinctions in the central Pacific. In *Extinction Rates*. John H. Lawton and Robert M. May, eds. Oxford: Oxford University Press, 75–87.

Pinker, Steven. 1994. *The Language Instinct: How the Mind Creates Language*. New York: William Morrow & Co.

Plenderleith, Kristina. 1999. The role of traditional farmers in creating and conserving agro-biodiversity. In *Cultural and Spiritual Values of Biodiversity: A Complementary Contribution to the Global Biodiversity Assessment*. Darrell Addison Posey, ed. London: Intermediate Technologies and United Nations Environment Programme, 287–291.

Posey, Darrell Addison, ed. 1999a. *Cultural and Spiritual Values of Biodiversity: A Complementary Contribution to the Global Biodiversity Assessment*. London: Intermediate Technologies Publications and United Nations Environment Programme.

Posey, Darrell Addison. 1999b. Introduction: Nature and culture—The inextricable link. In *Cultural and Spiritual Values of Biodiversity: A Complementary Contribution to the Global Biodiversity Assessment*. Darrell Addison Posey, ed. London: Intermediate Technologies and United Nations Environment Programme, 3–16.

Pottle, Frederick A., ed. 1950. *Boswell's London Journal, 1762–1763*. New York: McGraw-Hill.

Pottle, Frederick A., ed. 1953. *Boswell on the Grand Tour: Germany and Switzerland, 1764*. New York: McGraw-Hill.

Pottle, Frederick A. 1966. *James Boswell: The Earlier Years*. New York: McGraw-Hill.

Price, Glanville. 1984. *The Languages of Britain*. London: Edward Arnold.

Quine, W. V. 1969. Natural kinds. In *Ontological Relativity and Other Essays*. New York and London: Columbia University Press, 114–138.

Raven, Peter H. 1986. Personal communication to editors of *World Resources 1986*. World Resources Institute and International Institute for Environment and Development.

Raven, Peter H. 1987. The scope of the plant conservation problem world-wide. In *Botanic Gardens and the World Conservation Strategy*. D. Bramwell, O. Hamann, V. Heywood, and H. Synge, eds. London: Academic Press, 19–29.

Raven, Peter H. 1988. Biological resources and global stability. In *Evolution and Coadaptation in Biotic Communities*. S. Kawano, J.H. Connell, and H. Hidaka, eds. Tokyo: University of Tokyo Press, 3–27.

Reid, Walter V. 1992. How many species will there be? In *Tropical Deforestation and Species Extinction*. T. C. Whitmore and J. A. Sayer, eds. London: Chapman & Hall, 55–74.

Reid, Walter V., and Kenton R. Miller. 1989. *Keeping Options Alive: The Scientific Basis for Conserving Biodiversity*. Washington, D.C.: World Resources Institute.

Ribeiro, Darcy. 1957. *Culturas e Linguas Indigenas do Brasil*. Separata de Educacaõ e Cieñcias Socais no. 6. Rio de Janeiro: Centro Brasileiro de Pesquisas Educacionais.

Ribot, Théodule-Armand. 1881. *Les Maladies de la Mémoire*. Paris: Germer Baillière.

Robb, John. 1993. A social prehistory of European languages. *Antiquity* 67, 747–760.

Robson, G. C. 1928. *The Species Problem: An Introduction to the Study of Evolutionary Divergence in Natural Populations*. Edinburgh: Oliver & Boyd.

Rolston, Holmes, III. 1985. Duties to endangered species. *BioScience* 35(11), 718–726.

Rolston, Holmes, III. 1993. God and endangered species. In *Ethics, Religion, and Biodiversity: Relations Between Conservation and Cultural Values*. Lawrence S. Hamilton, ed. Cambridge, U.K.: The White Horse Press, 40–64.

Rosch, Eleanor. 1978. Principles of categorization. In *Cognition and Categorization*. Eleanor Rosch and Barbara B. Lloyd, eds. Hillsdale, N.J.: Lawrence Erlbaum Associates, 27–48.

Rouchdy, Aleya. 1989. Urban and non-urban Egyptian Nubian: Is there a reduction in language skill. In *Investigating Obsolescence: Studies in Language Contraction and Death*. Nancy D. Dorian, ed. Cambridge, U.K.: Cambridge University Press, 259–268.

Rowley, John. 1999. Wake up call. *People & the Planet* 8(3), 31.

Ruhlen, Merritt. 1991. *A Guide to the World's Languages. Volume 1: Classification*. Revised ed. Stanford, Calif.: Stanford University Press.

Ruhlen, Merritt. 1994. *The Origin of Language: Tracing the Evolution of the Mother Tongue*. New York: John Wiley & Sons.

Ruse, Michael. 1979. *Sociobiology: Sense or Nonsense?* Dordrecht: Reidel.

Schiller, Herbert. 1976. *Communication and Cultural Domination*. White Plains, N.Y.: International Arts and Sciences Press.

Schonewald, Christine M. 2001. Introduction to boundary space. *Complexity* 6:2, 41–57.

Sepkoski, John J., Jr. 1992. Phylogenetic and ecologic patterns in the Phanerozoic history of marine biodiversity. In *Systematics, Ecology, and the Biodiversity Crisis*. Niles Eldredge, ed. New York: Columbia University Press, 77–100.

Sepkoski, John J., Jr. 1999. Rates of speciation in the fossil record. In *Evolution of Biological Diversity*. Anne E. Magurran and Robert M. May, eds. Oxford: Oxford University Press, 260–282.

Shankar, Darshan. 1999. Cultural and political dimensions of bioprospecting. In *Cultural and Spiritual Values of Biodiversity: A Complementary Contribution to the Global Biodiversity Assessment*. Darrell Addison Posey, ed. London: Intermediate Technologies and United Nations Environment Programme, 534–535.

Shrestha, T. B. 1999. *Nepal Country Report on Biodiversity*. Kathmandu: IUCN–Nepal.

Shiva, Vandana. 1993. *Monocultures of the Mind: Perspectives on Biodiversity and Biotechnology*.

London, Atlantic Highlands, N.J., and Penang, Malaysia: Zed Books and Third World Network.

Simberloff, Daniel. 1986. Are we on the verge of a mass extinction in tropical rain forests? In *Dynamics of Extinction*. D. K. Elliott, ed. New York: John Wiley & Sons, 165–180.

Simpson, George Gaylord. 1949. *The Meaning of Evolution: A Study of the History of Life and of its Significance for Man*. New Haven, Conn.: Yale University Press.

Simpson, George Gaylord. 1961. *Principles of Animal Taxonomy*. New York: Columbia University Press.

Skutnabb-Kangas, Tove. 2000. *Linguistic Genocide in Education—Or Worldwide Diversity and Human Rights?* Mahwah, N.J.: Lawrence Erlbaum Associates.

Slikkerveer, L. Jan. 1999. Ethnoscience, 'TEK' and its application to conservation: introduction. In *Cultural and Spiritual Values of Biodiversity: A Complementary Contribution to the Global Biodiversity Assessment*. Darrell Addison Posey, ed. London: Intermediate Technologies and United Nations Environment Programme, 169–174, 176–177.

Smith, Anthony. 1980. *The Geopolitics of Information: How Western Culture Dominates the World*. New York: Oxford University Press.

Smith, F. D. M., R. M. May, R. Pellew, T. H. Johnson, and K. R. Walter. 1993a. Estimating extinction rates. *Nature* 364, 494–496.

Smith, F. D. M., R. M. May, R. Pellew, T. H. Johnson, and K. R. Walter. 1993b. How much do we know about the current extinction rate? *Trends in Ecology & Evolution* 8, 375–378.

Smith, Vera. 1999. Geological science: Where does in fit in ecosystem management? In *On the Frontiers of Conservation: Proceedings of the 10th Conference on Research and Resource Management in Parks and on Public Lands*. David Harmon, ed. Hancock, Mich.: The George Wright Society, 126–130.

Sneath, Peter H. A., and Robert R. Sokal. 1973. *Numerical Taxonomy: The Principles and Practice of Numerical Classification*. San Francisco: W. H. Freeman.

Snow, C. P. 1959. *The Two Cultures and the Scientific Revolution: The Rede Lecture, 1959*. New York: Cambridge University Press.

Soulé, Michael E. 1985. What is conservation biology? *BioScience* 35(11), 727–734.

Soulé, Michael E., ed. 1986. *Conservation Biology: The Science of Scarcity and Diversity*. Sunderland, Mass.: Sinauer.

Soulé, Michael E., and Gary Lease, eds. 1995. *Reinventing Nature? Responses to Postmodern Deconstruction*. Washington, D.C.: Island Press.

Soulé, Michael E., and Bruce A. Wilcox. 1980. Conservation biology: Its scope and challenge. In *Conservation Biology: An Evolutionary–Ecological Perspective*. Michael E. Soulé and Bruce A. Wilcox, eds. Sunderland, Mass.: Sinauer, 1–8.

Stattersfield, Alison J., Michael J. Crosby, Adrian J. Long, and David C. Wege. 1998. *Endemic Bird Areas of the World: Priorities for Biodiversity Conservation*. BirdLife Conservation Series no. 7. Cambridge, U.K.: BirdLife International.

Stendell, Rey C. 2000. Conservation biology: Achieving a balance. In *Links Between Cultures and Biodiversity: Proceedings of the Cultures and Biodiversity Congress 2000*. Xu Jianchu, ed. Kunming, Yunnan, China: Yunnan Science and Technology Press, 107–110.

Stewart, Charles, and Rosalind Shaw. 1994. Introduction: Problematizing syncretism. In *Syncretism / Anti-Syncretism: The Politics of Religious Synthesis*. Charles Stewart and Rosalind Shaw, eds. New York and London: Routledge, 20–21.

Stork, Nigel E. 1997. Measuring global biodiversity and its decline. In *Biodiversity II: Understanding and Protecting Our Biological Resources*. Marjorie L. Reaka-Kudla, Don E. Wilson, and Edward O. Wilson, eds. Washington, D.C.: Joseph Henry Press, 41–68.

Swadesh, Morris. 1971. *The Origin and Diversification of Language*. Chicago: Aldine Atherton.

Takacs, David. 1996. *The Idea of Biodiversity: Philosophies of Paradise*. Baltimore and London: Johns Hopkins University Press.

Tangley, Laura. 1986. Biological diversity goes public. *BioScience* 36(11), 708–715.

Taylor, Bron. 1997. Earthen spirituality or cultural genocide? Radical environmentalism's appropriation of Native American spirituality. *Religion* 27, 183–215.

Teilhard de Chardin, Pierre. 1956. The antiquity and expansion of human culture. In *Man's Role in Changing the Face of the Earth*. William L. Thomas, Jr., ed. Chicago: University of Chicago Press, 103–122.

Templeton, Alan R. 1989. The meaning of species and speciation: A genetic perspective. In *Speciation and its Consequences*. Daniel Otte and John A. Endler, eds. Sunderland, Mass.: Sinauer, 3–27.

Templeton, Alan R. 1991. Genetics and conservation biology. In *Species Conservation: A Population–Biological Approach*. A. Seitz and V. Loeschcke, eds. Basel: Birkhauser, 15–29.

Tenbruck, Friedrich H. 1991. The dream of a secular ecumene: The meaning and limits of policies of development. In *Global Culture: Nationalism, Globalization, and Modernity*. Mike Featherstone, ed. London: SAGE, 193–206.

The Nature Conservancy. 1975. *The Preservation of Natural Diversity: A Survey and Recommendations*. Washington, D.C.: The Nature Conservancy.

Thieberger, Nicholas. 1990. Language maintenance: Why bother? *Multilingua* 9(4), 333–358.

Thomas, William L., Jr., ed. 1956. *Man's Role in Changing the Face of the Earth*. Chicago: University of Chicago Press.

Thompson, John N. 1994. *The Coevolutionary Process*. Chicago and London: University of Chicago Press.

Thrupp, Lori Ann. 1999. Linking biodiversity and agriculture: Challenges and opportunities for sustainable food security. In *Cultural and Spiritual Values of Biodiversity: A Complementary Contribution to the Global Biodiversity Assessment*. Darrell Addison Posey, ed. London: Intermediate Technologies and United Nations Environment Programme, 316–320.

Todorov, Tzvetan. 1993. *On Human Diversity: Nationalism, Racism, and Exoticism in French Thought*. Catherine Porter, trans. Cambridge, Mass.: Harvard University Press.

Toledo, Victor M. 2001. Biocultural diversity and local power in Mexico: Challenging glob-alization. In *On Biocultural Diversity: Linking Language, Knowledge, and the Environment.* Luisa Maffi, ed. Washington, D.C.: Smithsonian Institution Press, 472–488.

Turner, B. L., II, William C. Clark, Robert Kates, John Richards, Jessica Mathews, and William Meyer, eds. 1993. *The Earth as Transformed by Human Action: Global and Regional Changes in the Biosphere Over the Past 300 Years.* Cambridge, U.K.: Cambridge University Press.

Upreti, Bishnu Raj. 2000. The effects of changing land-use systems in agricultural biodi-versity: Experiences & lessons from Nepal. In *Links Between Cultures and Biodiversity: Pro-ceedings of the Cultures and Biodiversity Congress 2000.* Xu Jianchu, ed. Kunming, Yunnan, China: Yunnan Science and Technology Press, 327–337.

Vitousek, Peter M., Paul R. Ehrlich, Anne H. Ehrlich, and Pamela A. Matson. 1986. Human appropriation of the products of photosynthesis. *BioScience* 36(6), 368–373.

Vitousek, Peter M., Harold A. Mooney, Jane Lubchenco, and Jerry M. Melillo. 1997. Human domination of Earth's ecosystems. *Science* 277, 494–499.

Wallace, Anthony F. C. 1966. *Religion: An Anthropological View.* New York: Random House.

Walter, Kerry S., and Harriet J. Gillett, eds. 1998. *1997 IUCN Red List of the Threatened Plants.* Gland, Switzerland, and Cambridge, U.K.: IUCN.

Watson, Seosamh. 1989. Scottish and Irish Gaelic: The giant's bed-fellows. In *Investigating Obsolescence: Studies in Language Contraction and Death.* Nancy D. Dorian, ed. Cambridge, U.K.: Cambridge University Press, 41–59.

Whewell, William. 1840. *The Philosophy of the Inductive Sciences, Founded Upon Their History.* London: John W. Parker. Facsimile reprint. London: Routledge/Thoemmes Press, 1996.

Whitehead, Alfred North. 1925. *Science and the Modern World: Lowell Lectures, 1925.* New York: Macmillan.

Whorf, Benjamin Lee. 1956. *Language, Thought, and Reality.* Cambridge, Mass.: MIT Press.

Wilson, Edward O. 1975. *Sociobiology: The New Synthesis.* Cambridge, Mass., and London: The Belknap Press of Harvard University Press.

Wilson, Edward O. 1978. *On Human Nature.* Cambridge, Mass.: Harvard University Press.

Wilson, Edward O. 1984. *Biophilia.* Cambridge, Mass., and London: Harvard University Press.

Wilson, Edward O., ed. 1988a. *Biodiversity.* Washington, D.C.: National Academy Press.

Wilson, Edward O. 1988b. The current state of biological diversity. In *Biodiversity.* E. O. Wil-son, ed. Washington, D.C.: National Academy Press, 3–18.

Wilson, Edward O. 1989. Threats to biodiversity. *Scientific American* (September), 108–116.

Wilson, Edward O. 1992. *The Diversity of Life.* Cambridge, Mass.: The Belknap Press of Har-vard University Press.

Wilson, Edward O. 1994. *Naturalist.* Washington, D.C.: Island Press.

Wilson, Edward O. 1997. Introduction. In *Biodiversity II: Understanding and Protecting Our Bio-logical Resources.* Marjorie L. Reaka-Kudla, Don E. Wilson, and Edward O. Wilson, eds. Washington, D.C.: Joseph Henry Press, 1–3.

Wilson, Edward O. 1998. *Consilience: The Unity of Knowledge.* New York: Alfred A. Knopf.

Wilson, Edward O. 2000. The creation of biodiversity. In *Nature and Human Society: The Quest for a Sustainable World.* Peter H. Raven, ed. Washington, D.C.: National Academy Press, 22–29.

Wittbecker, Alan E. 1991. Recognizing primary cultures as independent nations and creating a framework for them. *Pan Ecology* 6(4), 1–12.

Wittgenstein, Ludwig. 1953. *Philosophical Investigations.* G. E. M. Anscombe, trans. New York: Macmillan.

Wolfram, Walt, and Natalie Schilling-Estes. 1995. Endangered dialects: A neglected situation in the endangerment canon. *Southwest Journal of Linguistics* 14(1/2), 167–181.

Wollock, Jeffrey. 2001. Linguistic diversity and biodiversity: Some implications for the language sciences. In *On Biocultural Diversity: Linking Language, Knowledge, and the Environment.* Luisa Maffi, ed. Washington, D.C.: Smithsonian Institution Press, 248–262.

World Resources Institute, IUCN, and United Nations Environment Programme. 1992. *Global Biodiversity Strategy: Guidelines for Action to Save, Study, and Use Earth's Biotic Wealth Sustainably and Equitably.* Washington, D.C.: WRI, IUCN, and UNEP.

World Resources Institute, United Nations Environment Programme, United Nations Development Programme, and The World Bank. 1998. *World Resources 1998–99.* New York: Oxford University Press.

Wurm, Stephen. 1991. Language death and disappearance: Causes and consequences. In *Endangered Languages.* Robert H. Robins and Eugenius Uhlenbeck, eds. Oxford and New York: Berg, 1–18.

Zent, Stanford. 2001. Acculturation and ethnobotanical knowledge loss among the Piaroa of Venezuela: Demonstration of a quantitative method for the empirical study of traditional environmental knowledge change. In *On Biocultural Diversity: Linking Language, Knowledge, and the Environment.* Luisa Maffi, ed. Washington, D.C.: Smithsonian Institution Press, 190–211.

Zepeda, Ofelia, and Jane H. Hill. 1991. The condition of Native American languages in the United States. In *Endangered Languages.* Robert H. Robins and Eugenius Uhlenbeck, eds. Oxford and New York: Berg, 135–155.

INDEX

Numbers in italics indicate pages with figures.